"Parents and grandparents shouldn't miss *The Enchanted Hour*, but more important, we all need to heed this delightful book's wise advice: Please read to your kids. It's not the school's job to get our kids reading, it's our job—and it's a wonderful, magical act of love and caring."   —James Patterson, international bestselling author

"Gurdon's book is truly a gift. . . . The benefits are so blindingly clear that you will get to the end of *The Enchanted Hour* and wonder, "Why don't we do this more often?"   —*Washington Examiner*

"To develop a true passion of reading, it's important to consume books with your inner and outer voice. . . . After reading Mrs. Gurdon's wonderful book, I hope you'll want to do the same thing with your family and friends, too."   —*Washington Times*

"This thoughtful book draws together a wealth of neuroscience and behavioral research, interviews with experts, and telling anecdotes to make its case. Advocacy writing of the highest order."

—*Toronto Star*

"Gurdon is a reading-aloud tub thumper"   —*The Guardian* (London)

"As soon as I began to read, I was filled with that kind of engrossed blossoming that happens somewhere inside of you when you start a really nourishing book."   —*Sunday Times* (London)

"[*The Enchanted Hour*] is an inspirational account that will make you want to grab a book and share it with someone you love."

—*The Sun* (London)

"This book shines a light on the incredible power of reading aloud for everyone—not just children. . . . [*The Enchanted Hour* is] a joyous book that'll have you scrambling to the library."

—*Yours* magazine (London)

"A wide-ranging, erudite, engaging book." —*Saga Magazine* (London)

"Fascinating." —*Daily Express* (London)

"Reading out loud allows the reader and the audience to share the magical realism of storytelling." —*Catholic Herald* (London)

"Gurdon has a clear talent for weaving in the facts and figures alongside anecdotes and personal reflections." —*Parvati Magazine*

"The book really makes us think about the many benefits of reading aloud. . . . With gentle yet positive encouragement, this book shows the value of reading aloud at every age." —*Parents in Touch*

"This wonderful audiobook is a fully enjoyable listen. . . . [Gurdon's] pacing is excellent, and her ability to maintain listener attention is impressive. . . . An important and engaging audiobook." —*AudioFile* magazine

"This book is beautifully written and informative without being dry. . . . Gurdon uses real-life examples and gives book recommendations along the way." —Lackawanna Public Library

"You can feel [Gurdon's] passion and enthusiasm for the magic of books and reading, and I love how that came across in the writing. Although a nonfiction title, at times I found myself wrapped up in a cocoon whilst reading this, something I often only find in fiction reads; Meghan Cox Gurdon has perfectly captured the beauty of oral storytelling in *The Enchanted Hour*." —*Reading with Jade*

"This book is a must-read for anyone looking to improve their lives and the lives of their children. Highly recommend!" —*N. N. Light's Book Heaven*

"I highly recommend this book to any person, not just parents or teachers! You'd be amazed at the power we have to change our lives for the better just by using a book and our voice." —*Live, Read, and Prosper*

THE ENCHANTED HOUR

THE ENCHANTED HOUR

THE MIRACULOUS POWER OF READING
ALOUD IN THE AGE OF DISTRACTION

{ MEGHAN COX GURDON }

HARPER

NEW YORK · LONDON · TORONTO · SYDNEY

HARPER

FIRST HARPER PAPERBACKS EDITION PUBLISHED 2020.

Designed by Fritz Metsch

Library of Congress Cataloging-in-Publication Data has been applied for.

ISBN 978-0-06-256282-1 (pbk.)

23 24 25 26 27 LBC 9 8 7 6 5

For Hugo and the Chogen

The soul is contained in the human voice.

—JORGE LUIS BORGES

*Love doesn't just sit there, like a stone. It has to be made,
like bread: re-made all the time, made new.*

—URSULA K. LE GUIN

CONTENTS

AUTHOR'S NOTE

———•••••———

THIS BOOK HAD its origins in an article I wrote for the *Wall Street Journal* in the summer of 2015, "The Great Gift of Reading Aloud," which itself emerged from two decades of nightly reading to my children, and my dozen-plus years as the paper's children's book critic. A few lines from that piece and others for the *WSJ* survive in these pages, as do adapted excerpts from humorous family sketches that I wrote in the early 2000s.

All my source materials are listed in the notes at the end, and in the acknowledgments I name the people who generously shared with me their time and expertise. Any infelicities of data or interpretation will be mine, not theirs. The individuals who appear in these pages are all real, and I have faithfully recorded their words (sometimes eliding or tidying for clarity), but to protect privacy I've changed many of their names. Lines of dialogue are as close to the truth as memory, digital recordings, and contemporaneous notes permit. For simplicity, I often use the word *parent* to describe any given adult who reads to a child and trust that all the aunts, uncles, cousins, brothers, sisters, teachers, babysitters, and lovely next-door neighbors who read to children will understand that of course I mean them, too. Similarly, in the spirit of tradition (not to mention ease of reading) I use the pronoun *he* to describe any theoretical child.

When a book mixes memoir and advocacy with science, history, art, and literature, as this one does, some ideas, thinkers, and events are bound to go unremarked and uncelebrated. I hope the

reader will forgive these inescapable omissions. The same goes for
the books that I discuss and, especially, the lists of additional suggested titles at the end. These are not clinical, impartial, or complete guides to "correct" stories for reading aloud, but personal
favorites of mine and my children. Other families will prefer other
books, and why not? We don't live on Camazotz, the dark planet
in *A Wrinkle in Time*, where everyone has to conform. We can read
what appeals to us and skip what doesn't, and that's as it should be.
The important thing is to read—out loud.

INTRODUCTION

———

THE TIME WE spend reading aloud is like no other time. A miraculous alchemy takes place when one person reads to another, one that converts the ordinary stuff of life—a book, a voice, a place to sit, and a bit of time—into astonishing fuel for the heart, the mind, and the imagination.

"We let down our guard when someone we love is reading us a story," the novelist Kate DiCamillo once told me. "We exist together in a little patch of warmth and light."

She's right about that, and explorations in brain and behavioral science are beginning to yield thrilling insights into why. It's no coincidence that these discoveries are coming during a paradigm shift in the way we live. The technology that allows us to observe the inner workings of the human brain is of a piece with the same technology that baffles and addles and seems to be reshaping the brain. In a culture undergoing what's been called "the big disconnect," many of us are grappling with the effects of screens and devices, machines that enhance our lives and at the same time make it harder to concentrate and to retain what we've seen and read, and alarmingly easy to be only half present even with the people we love most. In this distracted age, we need to change our understanding of what reading aloud is, and what it can do. It is not just a simple, cozy, nostalgic pastime that can be taken up or dropped without consequence. It needs to be recognized as the dazzlingly transformative and even countercultural act that it is.

For babies and small children, with their fast-growing brains, there is simply nothing else like it. For that reason, I've devoted a substantial proportion of this book to the young. They respond in the most immediately consequential ways when someone reads to them, and as a result they are the subject of most research on the topic. As we shall see, listening to stories while looking at pictures stimulates children's deep brain networks, fostering their optimal cognitive development. Further, the companionable experience of shared reading cultivates empathy, dramatically accelerates young children's language acquisition, and vaults them ahead of their peers when they get to school. The rewards of early reading are astonishingly meaningful: toddlers who have lots of stories read to them turn into children who are more likely to enjoy strong relationships, sharper focus, and greater emotional resilience and self-mastery. The evidence has become so overwhelming that social scientists now consider read-aloud time one of the most important indicators of a child's prospects in life.

It would be a mistake, though, to relegate reading aloud solely to the realm of childhood. The deeply human exchange of one person reading to another is, in fact, human, which means that its pleasures and benefits are open to everyone. Teenagers and adults who are read to, or who do the reading themselves, may not win the same degree of scientific interest, but there's no question that they benefit, too, in ways intellectual, emotional, literary, and even spiritual. For frazzled adults in midlife, whose attention is yanked in a thousand directions, taking the time to read aloud can be like applying a soothing balm to the soul. For older adults in later life, its effects are so consoling and invigorating as to make it seem like a health tonic, or even a kind of medicine.

We have everything to gain and no time to waste. In the tech era, we can all benefit from what reading aloud supplies, but with children, the need is urgent. Many young people are spending as many as nine hours a day with screens. They are surrounded by technology—it informs their world, it absorbs their attention, it commandeers their hands and eyes—and they need the adults in their lives to read books to them not despite that but *because* of it.

In our cultural adaptation to the Internet, we have gained and we have lost. Reading aloud is a restorative that can replenish what technology leaches away. Where the screen tends to separate family members by sending each into his own private virtual reality, reading together draws people closer and unites them. Sitting with a book and a companion or two, we are transported to realms of imagination in warm physical proximity to one another. For children, contemplating the illustrations in picture books quietly and at length helps to inculcate the grammar of visual art in a way that really can't happen when the pictures are animated or morphing or jumping around. Where the infinite jostling possibilities of the touch screen make us feel scatterbrained, a story read aloud engages our minds in deep, sustained attention. The language of stories helps babies develop the linguistic scaffolding for early speech, and helps young children work toward fluency. When children are older, novels read aloud give them access to complex language and narratives that might otherwise be beyond them. The experience bathes children of all ages in torrents of words, images, and syntactical rhythms that they might not get anywhere else. It brings joy, engagement, and profound emotional connection for children, teens, adults—everybody. Reading out loud is probably the least expensive and most effective intervention we can make for the good of our families, and for the wider culture.

The Enchanted Hour is for anyone who loves books, stories, art, and language. It is for everyone who wants to give babies and toddlers the best possible start in life, everyone who cares about the tenderhearted middle-schooler and the vulnerable, inquisitive teenager, and for everyone who has yearned for an encounter with literature that breaks through what Virginia Woolf called "the cotton wool of daily life." It is for people who have never tried reading out loud. It is for people who've read aloud for years. Most of all, perhaps, this book is for everyone who has felt the dulling of emotional connection and the muddying of once-clear ideas and priorities in an era of noisy ephemera, technological enthrallment, and an overbearing news cycle.

In these pages, you will find enthrallment of a simpler kind. At its heart is the modest act of one person reading to another. It might be a teacher reading to a class, a mother reading to her children, a husband reading to his wife, or even a volunteer reading to a rescue dog. The act is simple, but its repercussions are complex and magnificent. In the chapters that follow, I will lay them out. We'll explore how sharing books enhances child development and why picture books are better than any tech or toy in giving young children what they need to flourish. We'll go back in time to an epoch when *all* reading was performed aloud, to gain a sense of the historical intertwining of voice and writing. We'll talk about audiobooks and podcasts. Then we'll explore the stupendous power of the spoken syllable to impart language, grammar, and syntax, and the ways that it can set the listener free from the confines of space and time. The reading voice has been a quiet source of entertainment by a thousand crackling fires and a bridge between generations. In a very real way it has also offered a ladder out of ignorance and an escape route from suffering and bondage—and it still does. It also helps listeners to discover what moves them, awakens an awareness of art and beauty, and equips young people to fulfill their potential as openhearted, curious, cultured adults.

My hope is that you will find the arguments, anecdotes, and research so exciting that you'll want to rush off to read aloud to the ones you love most. If that happens, I will have done my part in a great cultural relay race that began for me, as it does for many of us, when I was too young to know what was happening.

Was I three? Four? At the very edge of memory, I can hear my mother reading Stan and Jan Berenstain's *The Big Honey Hunt* and

Dr. Seuss's *Green Eggs and Ham*. My grandmother's voice is in there, too, reading *The Story About Ping*, by Marjorie Flack. The adults in my life stopped reading to me once I was able to read on my own, which is often what happens (and is cause for regret, as we shall see), and then I grew up and the subject went away.

I didn't give a thought to reading aloud one way or another for decades, though the idea of it, the beauty and importance of it, had evidently lodged itself in my consciousness. That slumbering idea woke up suddenly one evening when my then-fiancé and I went to a dinner party at the house of our friends Lisa and Kirk, who had a passel of small boys. During cocktails, as everyone was chatting, Lisa excused herself and disappeared upstairs. She was gone so long that eventually someone asked Kirk if there was something amiss. "Oh, no," he said. "She's just reading to the boys."

She's just reading to the boys. Any chagrin we might have felt at being stranded by our hostess was replaced, for me, by stunned admiration and a vow to do the same for my own children, if I ever had any. I would put reading aloud first for them, too.

So it was, twenty-four years ago, that when my husband and I arrived home from the hospital with our first baby, one single bright idea stood out in my bewildered postpartum mind like a neon sign in the fog. *I must read to this baby.* The moment the front door of our apartment clicked shut behind us, I carried the infant to a rocking chair and picked up a book of fairy tales. It was all very new, very strange, and very disorienting. I propped the book open and began to read.

"'Once upon a time,'" I told the baby, Molly, "'there lived a widower who had one daughter. For his second wife, he chose a widow who had two daughters. All had very jealous natures, which was unfortunate for the gentleman's daughter, because they made her stay at home and do all the hard work while they put on their finest dresses and went to garden parties . . .'"

Hot summer sun was slanting though the window. My voice

sounded false and stagey in my ears. The baby did not seem to be aware of what was happening.

"'The Prince was dancing a minuet with the elder of the step-sisters, when suddenly the music stopped and—'"

Was she even listening?

Was I supposed to show her the pictures?

Wait, was she asleep?

With a sudden sense of personal failure sharpened by exhaustion and the realization that the whole spectacle was absurd—what kind of maniac reads *Cinderella* to a newborn?—I felt my throat tighten and tears rush to my eyes.

It was a messy, inauspicious start to what would become our most beloved family ritual. *The Enchanted Hour* springs from those early tremulous days and the years that followed as Molly was joined by a brother, Paris, and three sisters: Violet, Phoebe, and Flora. I read to them for an hour or so every night—and I'm still reading today. In the wild time of their extreme youth, settling in after a long turbulent day for our evening retreat with books felt like reaching a life raft. Gratitude and relief would wash over me. We'd made it! *Now* we could relax. *Now* was the great part.

Was the time always enchanted? Certainly not. Reading aloud is often a sacrifice and sometimes a nuisance. Even for a zealot, it's not always easy to find the time or patience. There were nights when I felt half frantic with wanting to get everyone settled, and nights when the books we chose didn't satisfy any of us. Sometimes I squinted at the page through eyes smarting with fatigue. I read through head colds and sore throats and once, stupidly, right after oral surgery (and popped a stitch halfway through "How the Rhinoceros Got His Skin"). There were times I couldn't bear to pronounce every florid description and shortened paragraphs on the fly (sorry, Brian Jacques). There were books that so moved me that I cried, and made my listeners cry, too, because their mother was crying.

Shortly before Flora arrived in the fall of 2005, I became the children's book critic for the *Wall Street Journal*. Overnight, our house was flooded with new children's books. Fresh titles entered our reading rotation alongside classics and old favorites. For years I was knee-deep in children's books, hip-deep in children, and up to my neck in the parenting world.

Then came the first bittersweet departure. In early adolescence, Molly left our enchanted reading circle. A few years later, Paris did, too. Phoebe jumped the line and went third. It was Violet's decision to go a few years ago, at fifteen, that spurred me to write this book. As I was finishing it, I could see the first tentative signs in Flora that she was readying herself to move on. I'm still knee-deep in children's books, but soon I won't be reading them out loud, yet that sound you hear is not a suppressed sob of misery or wistfulness. It's the thwack of the baton as I pass it on to you.

Family life can be a hectic and flailing business. Sometimes it's a struggle to keep everyone afloat, let alone to haul them onto a read-aloud raft at bedtime. Yet it's an effort worth making, especially now that almost everyone's raft is tossing on a wide and often lonely sea of pixels. Young and old, we need what reading aloud has to offer. If I were Glinda the Good Witch from *The Wonderful World of Oz*, I'd wave my wand and bestow the gift on every household, everywhere. Since I am only myself, and wandless, I hope this book will cast the persuasive spell instead.

THE ENCHANTED HOUR

CHAPTER I

WHAT READING TO CHILDREN DOES TO THEIR BRAINS

———

In the great green room
There was a telephone
And a red balloon
And a picture of—
The cow jumping over the moon
And there were three little bears
    sitting on chairs . . .

—Margaret Wise Brown, *Goodnight Moon*

In 1947, Christian Dior introduced the New Look to women's fashion, Jackie Robinson signed with the Brooklyn Dodgers to become the first African American player in the major leagues, and a company called Harper & Brothers published a quiet little bedtime story entitled *Goodnight Moon*.

It was a consequential year! Dior's innovation sparked an exuberant postwar fashion renaissance, Robinson's dignity and athleticism inspired the world, and the quiet little picture book went on to become the single most cherished text of modern babyhood. Since its first appearance, *Goodnight Moon* has sold what I believe is known in the trade as a gazillion copies. Generations of children have listened as a grown-up reads the droll, crystalline verses by Margaret Wise Brown that describe a little rabbit's bedtime routine of bidding goodnight to the things in his room. Innumerable small fingers have touched Clement Hurd's color-saturated illustrations of the great green room, with its framed pictures, crackling fire, and wide windows with green-and-gold curtains. Innumerable pairs of eyes have lingered over the oddments that make the scene so delectable and distinctive: the rabbit's tiger-skin rug, the comb and the brush and the bowlful of mush, and the kittens playing with a skein of wool belonging to "the old lady whispering hush." As the pages turn and the evening ticks on, a tiny mouse moves from place to place and the bright moon rises in a starry sky outside.

When my children were small, *Goodnight Moon* was a big part of the evening ritual. I don't suppose we read it every night, but at this remove it seems as though we did. The cadence of the verse became as familiar and comforting to us as an old stuffed animal. The pictures, though, always had an element of novelty, because we were always looking for something new in them. At some point when Molly was a toddler, she and I invented a game that we called "quizzes." In the game, it was my job to try to stump her, and later her siblings, by asking them to seek and find obscure things in books such as *Goodnight Moon*. When the children were very small, it was enough of a challenge for them to spot "the bowl" or "the flames" or "the pair of slippers" in Clement Hurd's artwork. As they got older, I had to find more exotic objects and use more obscure language to keep the game going.

"Can you locate the two timepieces?" I might say. A small finger would shoot out to touch one clock on the mantelpiece and another on the rabbit's bedside table.

"What about the . . . andirons?"

That was a tricky one. I remember a long pause. At last I pointed to the mysterious objects (in the fireplace, holding up the burning logs), and tried something else.

"Who can identify the second moon?"

Out went another finger, pointing to a tiny crescent in the picture of the cow jumping through the night sky. At the time, I had no idea that our game was anything but a bit of fun, but we had, in fact, stumbled unknowing into the foothills of a mountain of pedagogical evidence. It turns out that getting young children to interact with texts, and talking with them about the pictures and stories as you go, hugely intensifies the benefits they get from the time you spend reading together. We'll look at this phenomenon in detail a bit later.

Our fondness for *Goodnight Moon* feels very particular to our family, in the way of the things we love in our own and our children's early years, but of course our attachment to it is just a tiny expression of its wider cultural significance. In the seven decades since the book was first published, its words and pictures have suffused childhood to such a degree that social scientists have come to use the phrase *"Goodnight Moon* time" to describe cozy parent-child time in the evenings—pajamas and tooth-brushing, reading out loud and tucking-in, and the general imparting of security and love before lights-out.

And why not? *Goodnight Moon* is perfect for the purpose. It is soothing. It is lulling. Millions of parents have turned to it at bedtime, not least, we can assume, because it helps to settle their children's minds into a state of placidity.

Well, appearances can deceive. A child listening to a storybook while looking at the pictures may seem placid, but beneath that tranquility, as we shall see, lies an incredible dynamism.

* * *

IF YOU WERE looking for the polar opposite of the great green room, you would not go wrong in identifying a certain chilled enclosure located deep in a research building attached to Cincinnati Children's Hospital Medical Center, on a hilltop in southwest Ohio. Having traveled along a gleaming hallway past an enormous cobalt wall fitted with video screens, and having passed through a series of blond wooden doors, you would come to an antechamber and two chambers divided by a plate-glass window. Let's call it the bland beige room.

Here there are no jolly pictures of a cow jumping over the moon, no fireplace, and no table lamp casting a pretty glow. Flashing lights and warning signs at the doorway hint at the seriousness of purpose of the place. In the first chamber, a desk stretches the width of the window, giving the technicians who operate its many monitors a good view of what's happening on the other side. Opposite, in the second chamber, is a kind of bed that is not the comfy sort occupied by the little rabbit in Margaret Wise Brown's story. This bed is narrow and designed so that its young occupant can be secured in position. Before a child lies down, he gets fitted with soft yellow earplugs and a set of headphones, and then is fastened in place with a strap. Once recumbent, his body is slid into the circular aperture of a resonance imaging machine, or MRI. There, on his back, surrounded by the racketing sounds of vibrating magnetic coils, he responds in the deepest portions of his brain to the sounds he hears through the headphones, and to the things his eyes see projected on a little mirror fixed above his face.

With a view of the child's small blanketed legs emerging from the machine, doctors—neurologists, radiologists, pediatricians, and researchers—can capture on their computers every flash of his thoughts, every evanescent streak that travels from one part of the brain to another.

These studies at the Cincinnati Children's Reading and Literacy Discovery Center are generating sensational insights into the effects of reading aloud on the developing brain. Among other discoveries, it seems that the thing we enthusiasts have long suspected is true: reading aloud really *is* a kind of magic elixir.

*   *   *

HALF A DOZEN miles away, rain was falling on the Cincinnati neighborhood of Oakley as babies, toddlers, and their caretakers pushed (and were pushed) into the warm, colorful interior of a children's bookstore. In contrast to the antiseptic hospital chamber, the walls here were covered with signed doodles and drawings left by visiting authors. The pictures didn't distract the rushing children, whose object was to get to a central area that had been cleared of its armchairs and sofas to accommodate a weekly combination of dance party and picture-book read-aloud.

"Look, there she is!" one mother said, directing her daughter's gaze to a purple-carpeted dais where manager and "story time lady" Sarah Jones was waiting with a guitar. Youthful and expressive, with her brown hair pulled back into a bun, "Miss Sarah" strummed a chord and beamed at the arriving hordes. A stunned-looking toddler in bib and striped pants stood beside her, face tipped up in guileless amazement. The child's older sister stood a short distance away in the same pose. Elsewhere toddlers knelt, squatted on their heels, or climbed onto grown-up laps as Jones began shifting chords to signal that the event was about to begin.

"Welcome, welcome every one," she sang, to the tune of "Twinkle, Twinkle, Little Star." Adult voices chimed in, and some of the children started to dance as Jones continued, "Now we're here to have some fun."

Fun was their purpose. Mine was observation. With my youngest child having hit double digits, it had been a while since I was immersed in the world of toddlers. I wanted to refresh my understanding of the ways they respond to stories in a group setting, and this was an ideal place to do it. Like me, the owner of the bookstore, Dr. John Hutton, has been reading aloud to his children for more than two decades. A pediatrician, he's also an assistant professor at the Cincinnati Children's Hospital and member of a triumvirate there using functional magnetic resonance imaging to study the effects of reading aloud on children's cognitive development. The scene spread out before us was like a year's worth of esoteric fMRI research brought to jumping, shouting life.

Still strumming her guitar, Jones said, "All right, friends! I'm so happy to see all of you here this morning!" She gave the strings a flourish, put down the instrument, and picked up a small stack of books. Leaning forward, she told the children that she was going to read them stories about sleepy farm animals, sleepy babies, and a sleepy solar system.

"Can you guess our theme today?" she asked.

This question brought a din of cries and yelps, all cheerful, none very coherent. The room by this time had just shy of thirty small children, and about the same number of parents, grandparents, and babysitters.

"First, we're going to start with a book called *Sleepy Solar System*," Jones said, displaying a front cover that showed three plump planets under a purple coverlet.

"Nighty-night!" someone shouted. A grandmother jogged a baby on her lap. One boy was still dancing. A few other children continued to churn around, but almost every face was directed toward the reader.

"'It's been a long, busy day in the starry Milky Way,'" Jones read aloud, stretching out the vowels. "'Sleepy, setting Sun calls out, "Bedtime, everyone."'" The whole scene, with children entranced, parents involved, rhyming songs, and picture-book stories, was a perfect feedback loop of emotional stimuli and literary nurturing.

Jones broke away from the story for a moment.

"Can you guys yawn along with the sleepy planets?"

"Yeearrrgh!" yawned all the children.

Dr. Hutton leaned over to me and said, sotto voce: "There are some days when Sarah is sick, or can't come in, and some other poor soul has to try to entertain these kids. We've had tantrums: 'I want Miss Sarah!'"

I laughed and turned back to watch the fray. In the book, the sleepy planets had settled down to rest in their nightcaps and curlers.

"Yay!" the audience cried, applauding.

*   *   *

A HUMAN BRAIN was floating on the computer screen when I went to see Dr. Hutton later in his office at the hospital. It was the image of a child's cerebral white matter, the nerve fibers of the brain that are sheathed in a protective material called myelin. The object on the screen was not shown as white, despite its name, but was rich with color. It resembled a glowing deepwater sea creature, a tangle of sensitive, delicate threads in psychedelic shades of azure, crimson, and lime green, all suspended in inky nothingness.

"This is like a wiring diagram," Dr. Hutton explained with a distinct lack of romance, pointing to the places where the threads crossed and converged. "Early experiences reinforce the connections and make the wiring stronger.

"Most of these things are going to develop normally, because they are genetically programmed. But the strength of the connections, the myelination, the wrapping of the nerves, is very responsive to stimulation. There's a maxim in neuroscience: 'Nerves that fire together, wire together.'"

With a click, Dr. Hutton removed the sea creature and pulled up gray slices of brains—it was less gory than it sounds—as seen from underneath. Deep in the core of each image I could see a small splash of scarlet, shaped something like a chili pepper. The red splashes varied in size from brain to brain. These images came from an exciting study that Dr. Hutton and his colleagues conducted a couple of years ago. They put a cohort of children between the ages of three and five through the scanner, one by one. (This is a lengthy process that starts with the patient preparation of each child to ensure that he will stay motionless inside the machine and can take up to forty-five minutes. "It's play-based," Dr. Hutton told me. "We say things like, 'You're going into a rocket ship now!' Or, 'You've got to sit really still and we're going to play the *statue game!*'") The researchers wanted to see what happened in the brains of these children when they heard age-appropriate stories read aloud. Which areas would be engaged? Would there be a difference in the neural response of children who had lots of experience being read to, compared with those who didn't?

The team discovered that the brains of young children whose parents read aloud to them often, and who had access to more children's books, had more robust activation than their peers. In other words, the brains of well-read-to preschoolers seemed more agile and receptive to narrative, suggesting that they had a greater capacity to process more of what they were hearing, and at faster speeds. This was the first study to show that the early home reading environment—which is to say, a child's access to books and frequency of shared reading with a grown-up—makes a quantifiable difference in brain function, and therefore, it stands to reason, in brain development. The researchers believe that because well-read-to children have greater experience with language and imagination during story time, they will enjoy a cognitive advantage over peers who have not. (A preschool teacher told me that she and her coworkers can always spot the well-read-to children: "They come in the morning and many of them will go straight to the books and say, 'Will you read me this book?' and they're like, looking for a lap," she said, at which point she stood up and waggled her bottom like a three-year-old trying to find a seat.)

The scarlet chili peppers that I saw in the pictures of these children's brains, indicating greater activation, were localized to the left posterior hemisphere of the brain, in a realm known as the parietal-temporal-occipital association cortex. This part of the brain is involved with processing multisensory information, in particular visual and auditory. This was the area that Dr. Hutton and his colleagues found to be more active in children who'd been read to the most. Strikingly, this particular study involved stories without pictures that children listened to through headphones, which suggested to the clinicians that activation in these visual processing areas represents imagination. Children with greater experience of being read to were, it seemed, better at summoning images in their mind's eye than young children who hadn't been exposed to lots of books and reading aloud.

Dr. Hutton's group has since published two additional first-of-their-kind fMRI-based papers involving preschoolers and the effects of reading aloud. One of them found that children who express more interest in listening to stories have greater activation in their cerebellum, the part of the brain that helps orchestrate skill refinement.

Well, you might ask, what of it? It's logical that a brain accustomed to certain stimuli is going to develop a greater capacity to handle those stimuli. Why does it matter? What difference does it make?

It matters because children's early years are a time of such intense formation. The young brain is plastic, adaptable, and growing like mad. In the first twelve months, a baby's brain doubles in size. By a child's third birthday, his brain has completed 85 percent of all the growth it will have. The sensitive period when synapses are forming for language and many other higher cognitive functions *peaks* when a child is two. By the end of the first five years, a child has passed through all the most rapid stages of development involving language, emotional control, vision, hearing, and habitual ways of responding. Early experiences and the firing and wiring of neurons create the architecture of a small child's mind, laying the pathways for future thought and reasoning.

Reading storybooks turns out to be an extraordinarily efficient and productive way to cause messages to zing from one part of the brain to another, creating and reinforcing those important neural connections. Reading aloud is so constructive in this regard, in fact, that in 2014 the American Academy of Pediatrics advised its 62,000 member doctors to recommend daily reading aloud to the parents and children they see in their medical practices. "Reading regularly with young children," the group's policy paper read, "stimulates optimal patterns of brain development, and strengthens parent-child relationships at a critical time in child development, which, in turn, builds language, literacy, and social-emotional skills that last a lifetime."

Optimal patterns of brain development! Stronger parent-child relationships! Skills that last a lifetime! If reading aloud were a pill, every child in the country would get a prescription.

Instead, we're giving them screens.

*   *   *

IT IS NO longer possible to think about children and their welfare without considering the effects of technology. Today, screens have entered even the most private and once-protected spheres of childhood. The consequences, both good and bad, are manifesting at every economic level and in every kind of family. According to one recent study, almost half of young children now have an electronic tablet or device of their own. Children ages eight and under are spending an average of nearly two and a half hours every day on screens. Averages do not, of course, reflect individual lives. Lots of very young children are spending far more time online than that. Older children are even more absorbed, with teens spending an average of six and a half hours a day using screen media, with more than a quarter of them on screens for eight or more hours—most of their nonschool waking time. And this is all before virtual reality goes mainstream.

When a child is watching a video story on a laptop or tablet, the act may appear indistinguishable from his looking at a story in a picture book while someone reads it to him. In each case, his eyes are taking in a series of illustrations, his ears are hearing the voice of a narrator, and his brain is making sense of what he sees and hears. But there is a powerful difference. These two means of receiving a story are in fact radically dissimilar, and given the ubiquity of screens and the amount of time kids spend on them, they diverge in ways that are profound, eloquent, and worrying.

Another first-of-its-kind study that Dr. Hutton and his colleagues conducted in 2017 suggests how and why this is the case. The researchers' goal this time was to expand their range of comparison by looking at brain activity when young children were listening to stories, when they were hearing stories while looking at illustrations—the classic picture-book experience—and finally when the same children were watching animated entertainment. We'll talk later about the value of listening on its own, without pictures, as you or I might do with an audiobook, but for the moment I'd ask you to consider the contrast between a storybook and a story video.

This time, the Cincinnati Children's team put twenty-eight children between the ages of three and five through the scanner, one by one, each time conducting a three-stage investigation of their brain activity. Each stage lasted for five minutes. To establish a baseline, the researchers collected images of what was happening when children were resting in the scanner looking up at the projection of a smiley-face emoji. Next, the image was removed and the experiment began. One by one, separated by intervals of rest, the preschoolers lay in the dark listening through headphones to stories with varying levels of visual stimulus. First came *The Sand Castle Contest* by Robert Munsch, read by the author. The children heard the story, but they saw no pictures. For stage two, the young subjects listened to another Munsch story, *Andrew's Loose Tooth*, also read by the author and accompanied this time by illustrations depicting scenes from the story. In the final phase, the children watched and listened to an animated version of Munsch's *The Fire Station*.

The objective was to see what happened in each circumstance in specific brain networks that support early literacy skills. The clinicians were looking at five areas: the cerebellum, the coral-shaped place at the base of the skull that is believed to support skill refinement; the default mode network, which is involved with internally directed processes such as introspection, creativity, and self-awareness; the visual imagery network, which involves higher-order visual and memory areas and is the brain's means of seeing pictures in the mind's eye; the semantic network, which is how the brain extracts the meaning of language; and the visual perception network, which supports the processing of visual stimuli. The clinicians measured the activation of these brain networks during each type of story, paying particular attention to how connected and synchronized they were.

The results were breathtaking. Dr. Hutton walked me through a chart displaying the team's preliminary conclusions. Rectangles in red showed the greatest activity of statistical significance, those in pink indicated somewhat less, and those in pale and dark blue showed the degree to which brain networks had faded or appeared idle.

We looked first at the data showing what happened when children were listening to a story without seeing anything. There was one red rectangle. "You're getting a little bit cooking, a few networks are up," Dr. Hutton said, "but really the one that stands out is the connection between the introspective areas, how does this relate to my life and understanding. There's not much in the visualization yet." This makes sense: young children have limited experience of the world and haven't built a large library of images, feelings, or memories to draw on.

He slid his finger over to the second column in the chart, the one showing neural activation as children looked at pictures while listening to a story.

"Bam!" he said. "All these networks are really firing and connecting with one another."

You didn't need a medical degree to see meaning in the serried stack of bright red boxes. When the children were listening to the story while looking at pictures, their brain networks were helping each other, reinforcing neural connections and strengthening their intellectual architecture, the delicate filaments of that floating sea creature.

Dr. Hutton was still pointing to the graph. "But then if you compare that with the video, everything kind of drops off," he said.

We sat in silence for a moment, looking at the third portion of the graph. All the red had turned blue.

"It's like the brain stops doing anything," I said.

"Except for the visual perception," he replied. "They're seeing the story and watching it, but nothing else is going on in terms of these higher-order brain networks that are involved with learning. What seems to be happening is the decoupling of vision, imagery, and language. The child is seeing the story and watching it, but not integrating this with other higher-order brain networks. The brain just doesn't have to do any work. In particular, imagination—supported by the default mode and imagery networks—falls off of a cliff."

"What are the implications of this?"

"In the behavior literature, it's clear that kids who have too much screen time can have deficits in different areas, like language, imagination, and attention," Dr. Hutton replied, his expression grave. "The years from three to five represent a formative stage of development. Too much screen time is a setup for atrophy, or underdevelopment of these higher-order brain networks. If what we know about brain plasticity is true, it will be harder for kids who grow up with underdeveloped networks to learn, to come up with their own ideas, to imagine what's going on in stories and connect it with their own lives, and they'll be much more dependent on stuff being fed to them, passively. I think it's a huge problem, and getting more complicated as screens become more portable. There is no natural barrier to use."

I looked back at the chart, which now seemed brutal in the comparisons it showed. "It's like all the color was stripped away," I said, "as if nothing is happening in their heads when they're watching the video."

"The lights are on," said Dr. Hutton, "but nobody's home."

There is a crucial reality to keep in mind here. The brains that seemed to flatline when their young owners were watching a video are the same brains that appeared to sparkle when presented with images and the sound of Robert Munsch reading his picture book. The Cincinnati Children's researchers have given this phenomenon a name: the Goldilocks effect. Like the bowls of porridge belonging to the three bears, one, audio, is "too cold" to get young children's brain networks engaging and integrating at optimal levels. Another, animation, is "too hot." Reading aloud from picture books seems "just right." Children have to do a bit of work to decode what they're hearing and seeing, which not only makes the experience engaging and fun but also helps reinforce the brain connections that will enable them to process harder and more complex stories as they get older.

What children don't get from one type of storytelling, in other words, they can get from another. But if the hours that children spend on screens do little or nothing to bolster their neurological development, as the research would seem to show, then it is all the more important that they spend time every day with an activity that does.

That is where the elixir of reading aloud comes in—and the sooner, the better. Children are only young for a little while, so this is not a matter for tomorrow, or sometime, or maybe never. It is something they need without delay. Reading aloud is not just a pleasant way to enjoy a story. It is a powerful counterweight to the pull of cultural and industrial forces that with stunning rapidity are reshaping infancy and childhood.

There's moral urgency, too.

*   *   *

IN 2015 A British political philosopher named Adam Swift infuriated parents across the English-speaking world by suggesting that people who read aloud to their children ought to reflect on the way they are "unfairly disadvantaging" *other* people's children. It was a puckish way of framing an uncomfortable truth, and, the Internet being what it is, irate correspondents deluged the University of Warwick professor with hate mail. Most of Swift's critics didn't take the trouble to read his original interview with the Australian Broadcasting Corporation, so they missed his most extraordinary assertion.

"The evidence shows," Swift had said, "that the difference between [children] who get bedtime stories and those who don't, the difference *in their life chances*, is bigger than the difference between those who get elite private schooling and those that don't." (Emphasis mine.)

Swift was using the phrase "bedtime stories" the way that others use *"Goodnight Moon* time," as academic shorthand for myriad informal behaviors that include reading aloud, such as "the talk at table, the family culture, the parenting styles, the inculcation of attitudes and values," as he put it.

Harvard political scientist Robert Putnam makes a similar case, calling *"Goodnight Moon* time" one of the most significant indicators of a child's academic prospects. In his book *Our Kids*, Putnam cites the work of Jane Waldfogel and Elizabeth Washbrook when he writes that "differences in parenting—especially maternal sensitivity and nurturing, but also provision of books, library visits, and the like—is the single most important factor explaining differences in school readiness between rich kids and poor kids, as measured by literacy, mathematics, and language test scores at age four."

Human development is cumulative. Each experience and each skill a person acquires informs the next. When it comes to reading aloud in childhood, the repercussions don't stop when kids get to school, or even when they reach adolescence. The ripples spread outward and onward into adulthood. This is true in a positive way when children are read to, and it's true in a lamentable one when they go without. A 2012 study found that children who enter kindergarten having had little or no *"Goodnight Moon* time" tend to lag other children by twelve to fourteen months in their language and prereading skills. Once at school, these children are just as likely as their peers to relish the fun and stimulation of story time, with the rhymes and humor and adventures and illustrations and all the rest of it. Yet in their innocence, they stand separated by the unforgiving math of a phenomenon called the word gap. A landmark study in the early 1990s uncovered stark differences in the number of words that children hear, or do not hear, depending on how they're raised: a gap of thirty million words by the age of three. (A 2017 study put the number at four million words by age of four; less of a gap, perhaps, but a chasm nonetheless.)

The implications matter—not just for individual children, but for our wider society—because early language and the cognitive and social skills associated with it are so closely linked with academic success. Recent inquiry has laid bare what might seem a counterintuitive link between the skills that children need to do well in English and the skills they need to do well in math. These two classroom subjects might seem on the surface to have little in common, but they have crucial, hidden points of connection.

When children struggle with math in middle school and the early high-school years, it turns out that the difficulty often lies less with numbers and numeracy than with words and reading. According to Dr. Candace Kendle, president and cofounder of Read Aloud 15 MINUTES, a national campaign to persuade parents to read daily to their children, "If you can't do fifth-grade reading problems, which are the first really analytical math problems that you encounter, if you can't process complex sentences, it is very hard to progress even into equation math and formula math, because you've missed this whole analytical process in the fifth grade.

"So when you think of how many kids aren't proficient or reading-ready in the fourth grade, it means that as a country we have immediately lost almost half of our potential science, technology, engineering, and mathematics workforce. It's frightening."

As CEO of a clinical research organization, Kendle experienced firsthand the difficulty in finding qualified young graduates to conduct lab work. "It's like forty-five percent of kids are not proficient," she said. "They may be able to read, but they're not proficient enough to do sophisticated analytical reading."

The numbers may be worse than Kendle thinks: a 2015 report found that 64 percent of US fourth-graders didn't meet the standards for proficient reading. If a fourth-grader can't read well today, it means that he wasn't up to snuff last year, either, nor probably the year before, when he was in second grade. The line of data and reasoning snakes backward through the elementary-school grades, back through kindergarten and nursery school, back to a child's earliest years. That precious early time is the starting place for academic deficits that may not become an obvious problem until high school.

Something like 20 percent—a fifth—of American teenagers leave high school functionally illiterate, meaning that they do not read and write well enough to navigate the working world. It is an awful way to start adult life. Eighty-five percent of kids who get into trouble with the law have poor literacy skills. Seventy percent of prisoners in state and federal institutions are in the same predicament, as are 43 percent of people living in poverty.

It's grim stuff. In this context, reading aloud to children becomes much more than a source of emotional and intellectual nourishment—though it is—more, even, than a developmental issue. Imagine how the world would look if every child had stories read aloud every night. As the picture-book creator and read-aloud advocate Rosemary Wells says, "We could narrow the achievement gap without spending another dime."

*   *   *

SO WHERE DO things stand? How many children are getting stories, and how many are not? We can find some answers in surveys of family reading habits that the publisher Scholastic runs every other year. In Scholastic's 2017 reading report, 56 percent of families reported that their babies had stories read aloud to them most days. The rates were higher for three-to-five-year-olds, with 62 percent of respondents saying their kids got the experience between five and seven times a week. Those numbers have been ticking up over time, which is great.

Flip the figures, though, and you will see a bleaker reality: 44 percent of US babies and toddlers and 38 percent of three-to-five-year-olds are *not* having stories read aloud to them often, or at all. In Great Britain, the numbers are actually dropping. A recent survey there found that the proportion of preschoolers getting a daily read-aloud has plummeted by almost 20 percent in the last five years, to just over half. (Nielsen Book Research, which conducted the survey, noted with alarm what appears to be an exact corresponding rise of 20 percent in the proportion of toddlers who watch videos online every day.) In short, millions of babies and young children are growing up, *right now*, at a disadvantage. Through no fault of their own, they are missing out on the emotional and intellectual nutrients that other kids get every day.

In a busy, distracted age, time and attention are in short supply. Finding a spare hour or even just a quiet fifteen minutes to read out loud can seem like an insuperable task. Even if parents aren't working long hours, or doing multiple jobs, it can be hard to summon the energy. And of course not every household conduces to moments of calm togetherness. Yet almost all parents have the chance to interact with their children at some point in the day. With a bit of ingenuity, that can become the moment to read together.

For some families, reading at breakfast while the baby is strapped into a high chair might be the best opportunity. It might be forty minutes with a toddler on the sofa before naptime, or ten minutes while father and daughter are waiting in a doctor's office. Reading might happen when young children are in the bath, or when everyone's taking the subway, or even over the phone to a child far away. It might mean converting a tedious half hour in an airport departure lounge to an enriching half hour of reading, or opening a book at the table while children are having an early supper of macaroni and cheese. It might be a full luxurious hour with the whole family every evening before bed. "Any time, any place," goes the read-aloud slogan of a literacy nonprofit in upstate New York, and that's exactly right.

As for what to read, books are obviously the ideal but in a pinch almost anything will do: a newspaper, a magazine, the laminated guide to emergency aircraft evacuation in the seat pocket in front of you. The British poet Roger McGough recalled semi-facetiously how his resourceful mother came up with reading materials during World War II: "Though books were scarce in those early years, mother made sure that I listened to a bedtime story every night. By the light of a burning factory or a crashed Messerschmitt she would read anything that came to hand: sauce bottle labels, the sides of cornflake packets. All tucked up warm and cozy, my favorite story was a tin of Ovaltine. How well I remember her voice even now: 'Sprinkle two or three heaped teaspoons of . . .'"

In that amusing recollection, you can hear the sense of enchantment, the ineffable magic in the mingling of a voice, a narrative, loving attention, and physical closeness. As the Cincinnati research suggests, and as we shall see in the chapters to come, the data show that wonderful things happen when we read aloud. Why this is so, exactly, is harder to say. At one level, it's a simple question of inputs: of language and consolation, of mutual attention and the pleasure of stories.

Yet there's a powerful quality of transcendence that raises reading aloud from the quotidian to the sublime. The experience is more than the sum of its parts. We can take it to pieces to see its beautiful and fascinating components—that's the work of this book—and yet at the same time, at its heart, we encounter a mystery. Like a biologist who has dissected the body of a songbird, we can see the pieces and how they fit together. We can identify wings and feet, beak and feathers. But we cannot see or hold the thing that made the bird so lovely to us: the grace of its flight and its warbling, fluting melody.

So it goes with reading aloud. Here is a reader, a book, a listener. The sound of the voice exists for a moment and then it vanishes. Like birdsong, it's gone—it is over. Yet it leaves traces of its passage in the imagination and memory of those who listen. There is incredible power in this fugitive exchange.

The story of humankind is the story of the human voice, telling stories. In reading aloud, we draw from an ancient wellspring of happiness that predates the written word. Oral storytelling has sustained and refreshed humankind since the far-off days of the distant past. So that's where we're going next.

CHAPTER 2

WHERE IT ALL BEGAN

Once Upon a Time in the Ancient World

———

Sing in me, Muse, and through me tell the story
of that man skilled in all ways of contending,
the wanderer, harried for years on end,
after he plundered the stronghold
on the proud height of Troy.

—opening lines of Homer's *Odyssey*, translated by Robert Fitzgerald

At the British Museum in London, down a long string of galleries filled with Greek antiquities, there is a glass case that contains a glossy black-and-ocher amphora, resembling a jug or vase. The object was made by a craftsman in Athens sometime early in the Golden Age, around 490–480 BC, and it's decorated with a figure on either side. The first is a musician in long skirts and a checkered tunic shown in full-length profile. We seem to have caught him just as he blows into a reed instrument.

On the other side, a man in pleated robes stands in a position of relaxed command, with one arm thrust out and resting on a tall wooden staff. The man's mouth is open, and if you look closely, you can see a tiny arc of text springing from his lips. Translated, the words read: "Once upon a time in Tiryns . . ."

This figure is a rhapsode, or "stitcher of songs," and a kind of living prefiguration of the act of reading aloud. In ancient Greece, a rhapsode did not read from a book, however. He *was* the book. His memory held, among other works, the two great epics of Homer,

The Iliad and *The Odyssey*. He would pull them from the shelf and read them aloud, so to speak, when he recited them.

The Homeric tales, loved to this day, are terrific creations. They brim with action, drama, stealth, deceit (and with manifestations of honor and dishonor so distinct from our own as to seem bizarre). *The Iliad* encompasses the ten years of the Trojan War, when the massed armies of the Greek kingdoms besieged the walled city of Troy. In its verses we meet sulky, ferocious Achilles, noble Prince Hector, handsome Paris, and lovely Helen. The second great Homeric tale, *The Odyssey,* follows Odysseus, wiliest of the Greeks, over the ten years it takes him after the conquest of Troy to reach his home island of Ithaca and his clever, long-suffering wife, Penelope. During his travels, Odysseus contends with mutinous crewmen, the erotic temptations of Circe and Calypso, and monsters such as the man-eating cyclops Polyphemus and the homicidal Sirens. At one point, Odysseus also has to wrest his men free of the addictive, obliterating pleasures of the lotus flower.

Today, if you pick up a printed and bound copy of *The Iliad* or *The Odyssey*, what you may notice first is not the richness of the storytelling but the sheer size of the thing. The epics are long and sprawling, and though they are strewn with mnemonic devices that would work as mental bookmarks for the would-be memorizer—vivid phrases and epithets such as "gray-eyed Athena," or "Zeus who wields the aegis"—it is still incredible to think that once upon a time people commited them to memory. Not only would a good rhapsode have both stories stored in his head, but he would be able to pick up either tale at any point and recite onward without a hitch. This is mastery of a sort that has become foreign to most modern people. With schools having largely withdrawn from the practice of making students memorize poetry, few of us today have anything approaching the interior resources of a rhapsode. You might argue that we don't need them: books are inexpensive and widely available, and we can use the Internet to look up pieces of writing that we may have forgotten or that we want to read. The rhapsodes themselves were obsolete long before the digital age was a glimmer in the eye of the future. Still, though they've long since disappeared, their role in the ancient world is a reminder that in reading aloud, we are taking part in one of the oldest and grandest traditions of humankind. Indeed, the long and rich lineage of reading aloud, as a type of oral storytelling, stretches back to the days before anything was written down.

*   *   *

LIKE A PERSON today who picks up a novel and reads out loud, rhapsodes were in the business of transmitting, not inventing. The opening words of *The Odyssey*—"Sing in me, Muse, and through me tell the story"—make this clear: The storyteller is acknowledging at the start that the tale he tells is not his own, and that he hopes for divine assistance in telling it well. You or I may not take such precautions when we open a storybook and read the words, but, like a rhapsode, we too are serving as a kind of artistic medium. We are drawing upon a story not of our own creation, and the story travels through us—through the concentration of our faculties, the inflection of our voices, the warmth and presence of our bodies—to reach the listener.

It is a marvelous thing: simple, profound, and very, very ancient. What Salman Rushdie calls "the liquid tapestry" of storytelling is one of the great human universals. So far as we can tell, starting in Paleolithic times, in every place where there are or have been people, there has been narrative. Here is *Gilgamesh*, the Sumerian epic recorded on clay tablets in cuneiform script 1,500 years before Homer. Here are the *Mahabharata* and the *Ramayana*, vast Sanskrit poems describing events in the ninth and eighth centuries BC. Here too is the thousand-year-old Anglo-Saxon legend *Beowulf*, the Icelandic *Völsunga* saga, the Malian epic *Sundiata*, the Welsh *Mabinogion*, the Persian, Egyptian, and Mesopotamian ferment of *The Thousand*

and One Nights, and the nineteenth-century Finnish and Karelian epic the *Kalevala*. This list is necessarily partial.

Once upon a time, none of these stories had yet been fixed on a page (or a clay tablet), but were carried in the physical bodies of the people who committed them to memory. Long before Johannes Gutenberg and his printing press, and a thousand years before cloistered monks and their illuminated manuscripts, the principal storage facility for history, poetry, and folktales was the human head. And the chief means of transmitting that cultural wealth, from generation to generation, was the human voice.

In ancient Greece, the voices belonged to rhapsodes; in ancient India, to charioteer bards called sutas. Elsewhere were skalds (Nordic history poets) and rakugoka (Japanese storytellers), along with the jongleurs, minstrels, and troubadours of medieval Europe. Shamans passed on the stories of tribal people native to North America. In West Africa there was, and is, an itinerant class of griots, the traveling tale-tellers and musicians who have been called living archives.

Even as human societies confided their stories and histories to print, people continued to rely on the voice to make sense of what was written. Until the tenth century AD, in fact, writing was not something to take in through the eyes and consider in silence with the mind. Rather, it was a mechanism for a kind of reverse dictation. To read at all was to read out loud. Sumerian cuneiform, Egyptian hieroglyphics, manuscripts in Aramaic, Arabic, and Hebrew, the illuminated Christian Gospels, the Talmud, the Koran—with these forms and collections of writing came the expectation that a person would read them out loud and would, in a manner of speaking, conjure their reality. In his book *A History of Reading*, Alberto Manguel points out that Aramaic and Hebrew, the "primordial" languages of the Bible, draw no distinction between reading and speaking. The same word stands for both. Buddhism and Hinduism also give an exalted place to the spoken word. What, if not reading aloud, are the guided meditations of Buddhism? What else is happening when, at the spring festival of Rama Navami, Hindus listen to readings from the *Ramayana*? (As a late-nineteenth-century Anglican visitor to India marveled, "Much merit is supposed to be derived even from the hearing of it.")

Silent reading of the sort we practice with our books and laptops and cellphones was once considered outlandish, a mark of eccentricity. Plutarch writes of the way that Alexander the Great perplexed his soldiers, around 330 BC, by reading without utterance a letter he had received from his mother. The men's confusion hints at the rarity of the spectacle. Six hundred years later, Augustine of Hippo witnessed the Milanese bishop (and fellow future saint) Ambrose contemplating a manuscript in his cell. Augustine was amazed by the old man's peculiar technique. "When he read," Augustine marveled in his *Confessions*, "his eyes scanned the page and his heart sought out the meaning, but his voice was silent and his tongue was still."

For Augustine, as Alberto Manguel observes, "the spoken word was an intricate part of the text itself." We don't think that way now. For us, the *written* word has the real weight and gravitas. We joke: "It must be true, I saw it on the Internet," in an echo of the old line about the sanctity of print.

Yet as Dante observed, speech—the words we say, the pauses between them, and our inflection—is our native language. Writing is the crystallizing of liquid thought and speech, and therefore a kind of translation. When a girl in modern times listens to her mother or father read an abridged version of *The Iliad* or *The Odyssey*, in a curious way she is hearing Homer translated at least four times over. What began as spoken Greek became written Greek, which was translated into written English, and then, in a final transformation, was freed from the page and set loose in the air as spoken English.

The liberated word is a marvelous thing, because almost everyone can take it in without making any effort. The reader (or troubadour or skald) must expend some energy to present a text, but for the auditor there is no requirement other than to supply his attention. If you are listening, it makes no difference whether you're hearing a narrative taken from someone's memory, à la rhapsode, or a text read from the page of a book. In either case, you are getting the story in its living, spoken form.

Speech is the way language comes to all of us first, in the beginning. We hear it. Then we speak it. Only later and with considerable study will we learn to read and write it. As we know from the troubling statistics about American high-school graduates, not everyone reaches proficiency. According to the United Nations, about 14 percent of the world's adult population is unable to read. Yet if illiteracy is a barrier to economic advancement, it has never been an obstacle to the enjoyment of storytelling, either for unlettered adults or for children who may not be old enough, deft enough, or motivated enough to decode a printed text. For them, as for the men and women in fourteenth-century England who gathered to hear Chaucer read from his *Canterbury Tales*, or the villagers in eighteenth-century Mali who came running when the griot turned up, the oral tradition offers sanctuary and delight.

*   *   *

HOW DO STORIES begin? "Once upon a time," as the rhapsode on the amphora in the British Museum is shown to say. These are the real magic words, whether in that most familiar form, or in a variant heard in Indonesia, "It was . . . and it was not," or in the Jamaican way of opening stories, "Once it was a time, a very good time. Monkey chewed tobacco and he spit white lime . . ." Faster than a genie, the words conjure a portal and take us through it, transporting us from the here and now to the realm of storytelling, a place that may be fantastic or realistic or some combination of the two. "Who can catalog the myriad ways that human beings use to signal, 'Now, I am telling you a story'?" asks the literary critic Laura Miller. "The speaker leaves off ordinary talk, the listener recalibrates her attention, and both enter into a relationship older than the memory of our race. A story takes us, for a while, out of time and the particularities of our own existence. The initiation into this ritual might come as a pause, a change of tone . . . [and it] tells us that a special kind of language, the language of story, has begun."

One autumn afternoon not so long ago, the initiation into this ritual took the form of a school principal handing a microphone to a twelve-year-old boy who was wearing shorts, a T-shirt, and turquoise socks. Throughout the morning, the fifth-grader and the other pupils at an all-boys school in suburban Maryland had competed in preliminary rounds of a tournament called "the Bard." I'd watched the younger boys taking turns reciting poetry in the sunshine at the foot of a small stone amphitheater. Some of their pieces were short: at least two boys confined themselves to the six brief lines of Alfred, Lord Tennyson's "The Eagle"; but I'd also seen a fourth-grader reciting the G. K. Chesterton poem "Lepanto" for what seemed like a good ten minutes. The fifty-odd little scholars fanned out on the steps had surprised me with their attentiveness, though the vigilance of their teachers probably also had a pacifying effect.

Now everyone had jammed into the gym to hear the finalists. You can imagine the roar. Five hundred boys between the ages of nine and eighteen were jostling for seating. It is perhaps harder to imagine (but I promise it happened) the total silence that fell when the principal put the microphone into the hand of that fifth-grader. This boy had won the middle-school preliminary round. Now he began to recount one of the most poignant farewells in literature, from book 6 of *The Iliad*.

"'Hector hurried from the house when she had done speaking,

and went down the streets by the same way that he had come,'"
the boy recited, loud and clear. "'When he had gone through the
city and had reached the Scaean gates through which he would go
out on to the plain, his wife came running towards him.'"

Hector does not know, but suspects, that it may be the last time
he sees his wife, Andromache, and their infant son, Astyanax, a
child "lovely as a star." Andromache begs her husband not to go:
"Your valor will bring you to destruction; think on your infant
son, and on my hapless self who ere long shall be your widow."

Hector can foresee the awful consequences, but he has to go
into battle; honor demands it. In his final moments with his wife,
he tries to steel them both for the fates they can expect at the hands
of the Greeks: for him, a violent death; for her, enslavement in an
invader's household. He hopes to die first, and be spared the sight
of her suffering: "May I lie dead under the barrow that is heaped
over my body ere I hear your cry as they carry you into bondage."

Moved by his own words, Hector reaches for his son but Astyanax, frightened by his father's shining armor and his helmet with
its nodding horsehair plume, hides his face in the bosom of his
nurse.

"'His father and mother laughed to see him,'" declaimed the boy
in the gym, pacing a little, "'but Hector took the helmet from his
head and laid it all gleaming upon the ground. Then he took his
darling child, kissed him, and dandled him in his arms, praying over
him the while to Jove and to all the gods.'"

I looked across the gym. Some of the younger boys were fidgeting in their seats, but no one talked or goofed around.

"Jove," Hector cries, "grant that this my child may be even as
myself, chief among the Trojans; let him be not less excellent in
strength, and let him rule Ilius [Troy] with his might. Then may
one say of him as he comes from battle, 'The son is far better
than the father. May he bring back the blood-stained spoils of him
whom he has laid low, and let his mother's heart be glad.'"

A moment later the scene ended. Feet pounded, hands clapped, throats vibrated, and the bleachers thundered. Through a skinny medium in surprising socks, Homer's poetry and pathos had reached, as if with a giant hand, from the foreign, distant past to hold these modern boys in its grip.

You might think there could be no real continuity between a bearded rhapsode in classical Greece two-and-a-half millennia ago and a boy in a modern gym. Yet there *is* a direct line, for reading aloud—its capacity to enthrall and enrich—is not in the look of the thing. It is in the telling.

*   *   *

ON A SEPTEMBER morning in 1858, in the town of Harrogate, North Yorkshire, a man sat weeping as Charles Dickens read aloud to a large audience from his novel *Dombey and Son*. The death of six-year-old Paul Dombey seemed to have been the poor fellow's undoing, and his sorrow was not lost on the author. "After crying a good deal without hiding it," Dickens wrote to his sister-in-law, Georgy, "he covered his face with both his hands and laid it down on the back of the seat before him, and really shook with emotion."

In the audience at the same reading, Dickens reported the presence of "a remarkably good fellow of thirty or so," who was so tickled by the comical character of Toots that "he *could not* compose himself at all, but laughed until he sat wiping his eyes with his handkerchief. And whenever he felt Toots coming again, he began to laugh and wipe his eyes afresh, and when he came, he gave a kind of cry, as if it were too much for him."

Novels including *Oliver Twist*, *Hard Times*, and *David Copperfield* had made Charles Dickens the J. K. Rowling of the nineteenth century. His readers in the United States were so worried about the fate of poor little Nell in *The Old Curiosity Shop* that crowds mobbed the docks in New York for the latest installment of the tale, much as, a hundred and fifty years later, hordes would assemble outside bookstores to await the midnight release of the newest book in the Harry Potter series.

In the time of Dickens, reading aloud at home was very much a common household entertainment. The practice had become broadly accessible in Britain a hundred years earlier, with the spread of literacy and the increased availability of books and periodicals. As Abigail Williams writes in *The Social Life of Books*, "People shared their literature in very different ways: reading books together as a sedative, a performance, an accompaniment to handiwork, a means of whiling away a journey or a long dark evening. They saw reading as a pick-me-up and a dangerous influence, a source of improvement, a way to stave off boredom, and even as a health-giving substitute for the benefits of a walk in the open air."

From a commonplace diversion, reading out loud became fashionable. The burgeoning elocution movement drew aspirants from up and down the social scale—from bakers' apprentices to clergymen to cloistered noblewomen—all eager to learn to read with elegance and panache. A person could gain social credit by reading well, whereas, of course, reading poorly meant courting embarrassment. Elocution guides warned of the shame of speaking with monotony, "like an ignorant Boy, who understands not what he reads." (History records the discomfiture of the great Jane Austen, who suffered one evening in 1813 as her mother rushed inexpertly through a passage of *Pride and Prejudice*. "Though she perfectly understands the characters herself," Austen confided in a letter to her sister, "she cannot speak as they ought.")

For Dickens, reading out loud in company was sufficiently popular that he could count on his audience knowing what he meant when he began his presentations, as he did, with a request: "I would ask that you imagine you are with a small group of friends, assembled to hear a tale told." Ladies and gentlemen in large, crowded public places might ordinarily wish to guard their emotional expression. By urging them to think of themselves as "a small group of friends," Dickens was inviting his listeners to surrender to the story with all the sincerity and openheartedness they would risk in private. As we know from his letters and diaries, people gave themselves up altogether to the emotional transport that he offered. Dickens was a talented reader who spent weeks practicing his delivery before taking his stories on the road. In the manuscripts that he used for public readings, he made marginal notes to remind himself of the various tones to strike from passage to passage ("Cheerful . . . Stern . . . Pathos . . .") along with gestures ("Point . . . Shudder . . . Look Round in Terror . . .").

Charles Dickens had to be sure that his paying customers went away satisfied. He was doing the job for money, after all. For most of us, the stakes are not economic. Reading at home, for love, we don't have to render up the prose in quite such a polished way. Even without shudders and well-timed moments of pathos, it is astonishing how effective a good story read aloud can be.

* * *

WHILE DICKENS WAS entertaining audiences in England and Ireland, across the Atlantic settlers were fanning westward across the United States, and they took his writing with them. For many emigrant families, a good session of reading magazine serials aloud in the evenings offered diversion from the lonely, arduous work of busting sod and planting crops. The chief difficulty was getting hold of fresh copies of magazines that carried the stories of Dickens, Victor Hugo, and others. After the opening of the transcontinental railway in 1869, the transit of the mails became smoother, but people still couldn't predict when or if their longed-for parcels, letters, or publications would arrive. As one emigrant recalled: "We were never able to be sure that we would be able to understand the next chapters in serial stories, which were our delight. I remember being very engrossed in one of Charles Reade's novels, the heroine of which was cast on a desert island . . . the story was published in [a magazine called] *Every Saturday*, and at first it came weekly, but after we had become most deeply interested, five weeks passed during which not a single number was received, and we were left to imagine the sequel."

Laura Ingalls Wilder captures the sense of relief and transport that pioneer families got from these luxuries—as well as their frustrating rarity—in *The Long Winter*, which, like the other novels in the Little House series, draws on real events. A few days before Christmas 1880, in the small Dakota Territory town of De Smet, Laura's father returns from the post office with an armful of magazines and newspapers sent by well-wishers. Laura and her sisters, Mary and Carrie, are dying to dig into the tantalizing stuff, but there's work to do first. Instant gratification was not a feature of the Ingalls household.

"Come girls," Ma says, "put the bundle of *Youth's Companions* away. We must get out the washing while the weather's clear." By the time the girls finish their chores, it's too dark to read; the Ingalls are almost out of kerosene, and with the trains blocked by snow, there's no more to be had. The next day, Ma proposes delaying their reading enjoyment yet again by holding off on the magazines until Christmas.

After a moment Mary said, "I think it is a good idea. It will help us to learn self-denial."

"I don't want to," Laura said.

"Nobody does," said Mary. "But it is good for us."

Sometimes Laura did not even want to be good. But after another silent moment she said, "Well, if you and Mary want to, Ma, I will. It will give us something to look forward to for Christmas."

"What do you say about it, Carrie?" Ma asked, and in a small voice Carrie said, "I will, too, Ma."

Come Christmas, the girls have to finish all their work first, and only late in the day does the yearned-for moment arrive:

"You girls choose a story," Ma said. "And I will read it out loud, so we can all enjoy it together."

So, close together between the stove and the bright table, they listened to Ma's reading the story in her soft, clear voice. The story took them all far away from the stormy cold and dark. When she had finished that one, Ma read a second and a third. That was enough for one day; they must save some for another time.

Around the time that Ma Ingalls was portioning out those stories, some two thousand miles to the south and east Cuban immigrants were reviving in Florida a wonderful practice that the colonial Spanish had outlawed at home. In 1865, as Alberto Manguel recounts in *A History of Reading,* a poet and cigar-maker named Saturnino Martinez had founded a newspaper expressly for workers in the cigar trade. As Manguel explains, "Over the years, *La Aurora* published work by the major Cuban writers of the day, as well as translations of European authors such as Schiller and Chateaubriand, reviews of books and plays, and exposés of the tyranny of factory workers and of the workers' sufferings." There was a problem with the newspaper, though. Most of the cigar workers couldn't read it; the literacy rate among the working class was around 15 percent. So Martinez came up with the idea of organizing public readers who could read the paper out loud to the men as they worked. In 1866, the first of these lectors, as they were known, pulled up his chair in a Havana cigar factory and began to read the news. The cigar-rollers pitched in to pay his wages, and in return they received hours of intellectual diversion.

Alas, it was not to last. Six months later, the authorities cracked down. If illiterate workers could "read" the newspaper by listening to it, they might get dangerous ideas. The authorities repressed the practice, and in Cuba it seems to have died away.

When large numbers of Cubans moved to Florida in the 1870s, however, the fashion revived. The son of a lector who read aloud to cigar workers in Key West at the turn of the twentieth century recalled his father's daily routine: "In the mornings, he read international news directly from Cuban newspapers brought daily by boat from Havana. From noon until three in the afternoons, he read from a novel. He was expected to interpret the characters by imitating their voices, like an actor." (*The Count of Monte Cristo*, by Alexandre Dumas, was apparently a great favorite of the Cuban cigar men.) Rolling cigars by hand all day may be dull work, but at this remove there seems something almost romantic about it. An 1873 magazine sketch shows a row of mustachioed men, all in short-brimmed hats, seated at a wooden table making cigars while, behind them, a fellow in spectacles sits upright with his legs crossed, reading aloud from a hardback book. It is a scene of calm, order, and industry.

Five years after that etching appeared in print, there was a momentous evolution in the very idea of reading aloud. Using a hand-operated mechanical roller fitted with tinfoil, Thomas Edison made the first recording of the human voice. The original phonograph track was restored several years ago, and if you listen closely, beneath a rush of crackle and noise, you can hear a man (it may be Edison) performing a loud recitation of "Old Mother Hubbard" and "Mary Had a Little Lamb." The man's voice has an amused sound, and he laughs now and then, as if he knows himself to be play-acting. Edison was in fact amazed to find that his contraption worked. "I was never so taken aback in my life," he said; "I was always afraid of things that worked the first time."

If Edison had anticipated the significance of the moment, he might have chosen poetry of a more exalted type. Still, in a way it seems appropriate that the first lines recorded by machine should come from nursery rhymes. After all, the technology itself was in its infancy. When it grew up, it would change everything.

*   *   *

IN THE 1930S, a new kind of rhapsode entered private homes in the United States and Great Britain. These readers were inhuman in their untiring diligence with long texts because they were, in point of fact, machines that played "talking books" on records. For one group of people in particular, these readers arrived not a moment too soon.

After World War I, great numbers of men with grievous injuries returned home, many having been blinded in the horrific haze of chemical weaponry. A person who loses his sight in adulthood is not just plunged into darkness but thrust overnight into illiteracy. It was possible in theory for the blinded veterans to learn to read braille with their fingertips, but not so easy in practice. Braille, like any other second language, is much harder to learn in adulthood. The plight of these stranded men added urgency to a project already in motion.

"There was a real sense that these people coming back needed to be trained in some way, nourished spiritually and emotionally, in every sense you can think of," said Matthew Rubery, a professor at the University of London and author of *The Untold Story of the Talking Book*.

The first full-length recorded books included Bible readings (the Gospel According to John, in the plummy tones of a BBC announcer) and works of fiction (Joseph Conrad's *Typhoon* and Agatha Christie's *The Murder of Roger Ackroyd* among them). For the war-blind, talking books came as a huge relief. They were free at last to listen to what they liked. They no longer had to rely on relatives or volunteers who might intrude with their own opinions, read without style or skill, or censor passages they thought indelicate or unsuitable for the listening ear.

Not until the mid-1970s and the advent of recorded books for commuters, however, did audio recordings begin to make real commercial inroads in the wider culture. And for a long time, they carried a whiff of illegitimacy. An audiobook might include all the words that an author had written, the thinking went, but you couldn't call it a *real* book. You couldn't claim to have *read* it. Listening was like cheating, wasn't it? Audiobook enthusiasts were inclined to be apologetic and even a little ashamed if the question came up.

Today audiobooks are a $3.5 billion industry and a huge source of pleasure and education for millions of people. Yet the shift in acceptability has only just happened, and there is no doubt that we have technology to thank for it. When Matthew Rubery set out to write his history of audiobooks in 2010 and was looking for funding, he had a hard time getting senior scholars to vouch for him. The subject seemed too frivolous. A few years later, as he was finishing his manuscript, grant money came pouring in. It was a sign that audiobooks had completed their cultural conquest.

And why not? It's a marvel that for a small fee, or for free with a library card, the world's most gifted readers will deliver any book we like, straight into our heads. Like the Cuban cigar makers, we can vanish into literature or nonfiction even as our hands are drudging—or driving or hanging on to the guardrails of a tread-mill.

For busy readers, podcasts and audiobooks are a boon. For people who cannot see or read well, they're a godsend. Matthew Rubery told me: "So many people with dyslexia come up to me and talk about what audiobooks have meant to them by taking something they absolutely hated and turning it into a joy."

From Edison's scratchy wax recording came shellac records, then vinyl, then audiotape, then CDs, then streaming, and, as of 2018, a return to vinyl for a hipster resurgence of "spoken word" recordings. Most new cars now come fitted with technology that makes it easy to listen to smartphone audio. Future generations of children will never have to fumble to extract cassette tapes from their brittle, clear plastic cases, nor will their parents ever taste the insouciant terror of hurtling down the highway, one hand on the wheel while scrabbling with the other to slip a CD out of its case and get it into the player, all the while reassuring the restive passengers, "Hang on, the story is coming—"

I have warm memories of those days, maddening plastic cases notwithstanding. On long journeys, my children and I would listen to Peter Dennis reading A. A. Milne's *Winnie-the-Pooh* and Martin Jarvis reading Kenneth Grahame's *The Wind in the Willows*. So loyal were my children to the recordings that when I brought out the books, they insisted that I read in the exact cadences of Peter Dennis and Martin Jarvis. To this day, my children describe a moment of pathetic confusion as being "alone in the moon," like Piglet, when he's running along carrying Eeyore's birthday balloon and it pops in his face.

Audiobooks and podcasts can be so good, so absorbing, that it may seem churlish to suggest that they could be in any way inferior to books read aloud in person. You might argue that they are, in fact, superior: any given amateur may fumble with language or phrasing, whereas a skilled narrator like Juliet Stevenson or Jim Dale will never falter. A superb audio recording is a work of art, a creation not so different in certain respects from a marble bust or a portrait in oils. Yet as with such works, the relationship between the adored and the adorer is one-sided. Unlike a live reader, a recording has no idea who is listening. It doesn't know or care what you feel. A recorded story does not spring to life in the moment, in that unpredictable, fugitive communion of voice, ears, and text. The machine won't stop to ask questions or to raise comparisons or engage you in any way. It will keep playing until the story ends, or you switch it off, or the battery runs down.

This is why, much as I myself love audiobooks, it seems to me that, outside the car, they must take a narrow second place to live readers—especially when children are involved. I was a little surprised to find that Matthew Rubery, the talking-book historian, agreed. He told me: "People who like audiobooks often try to defend them by saying that they're hearkening back to the ancient tradition of reading aloud. And of course, we have been listening to stories much longer than we have been reading them silently. But it's not really true. It's very different to, say, listen to Homer read by Derek Jacobi on your earbuds than to listen to a rhapsode in ancient times.

"I think it's that personal dimension," Rubery went on. "Even if it's somewhat impersonal—let's say a large group and a performer is reading to you—that performer is still going on cues from the audience, whether they're interested or yawning or whatever. Certainly, when I'm reading to my kids, I respond to them in a lot of ways. If I see they're bored, I'll change my approach. I'll do a voice differently, or emphasize things differently. That's the virtue of reading aloud."

It *is* the virtue of reading aloud, and it's what the best reciters, storytellers, and readers have done for millennia. Somehow, though, without thinking it through, in our modern culture we have drifted away from this rich tradition. Some children have stories read to them in the evenings, and that's marvelous. Some adults have stories read to them via headphones, and that's good, too. But more often than not, we are doing something else with our time. We are looking at screens, and we're scrolling—scrolling alone.

There is a scene in *The Odyssey* that brings to mind our present predicament. It comes early in the epic. Odysseus and his men have left Troy on their way home to Ithaca and have to make landfall on an island unknown to them. Expecting attack at any moment, Odysseus sends three emissaries to intercept the inhabitants. Far from greeting the strangers with violence, the tranquil island people offer the men a delicious food, the honey-sweet lotus. One taste of the stuff, Homer tells us, and the sailors "never cared to report, nor to return: they longed to stay forever, browsing on that native bloom, forgetful of their homeland."

Thus we might describe the effects of screen-based technology on us. Thus the toddler with an iPad, who cares for no other toy; thus the middle-schooler, indoors and online as the sun shines and sets unseen; thus the teenager behind a closed bedroom door, lost for hours in the wonder-worlds of her phone; thus the grown-up so preoccupied with Twitter that he burns the soup.

Among the lotus-eaters, Odysseus is quick to see the danger, and he has to resort to force to dislodge his captivated crewmen. He says: "I drove them, all three wailing, to the ships, tied them down under their rowing benches, and called the rest: 'All hands aboard; come, clear the beach and no one taste the Lotus, or you lose your hope of home.'"

Homer could not have anticipated our tech devices, but he captures their lotus-like allure. He also shows us a way out. If we are wise, we will drag ourselves and our families, wailing if necessary, to ships that wait to sail on what literary historian Maria Tatar calls "the ocean of stories." If we can ease off the lotus-eating, just for a little, we can clear our heads and return to a different kind of home. Humankind has flourished with the sharing of stories from its earliest days. In reviving the art of reading aloud, we can reclaim an old pleasure, one that has an amazing capacity to draw us closer to one another.

CHAPTER 3

READING TOGETHER
STRENGTHENS THE BONDS
OF LOVE

In the bright buzzing room
There was an iPad
And a kid playing Doom
And a screensaver of—
A bird launching over the moon.

—"Ann Droyd," *Goodnight iPad*

In 2011, the Apple tablet had been on the market for a year and a half when a pseudonymous wit published a "parody for the next generation" of Margaret Wise Brown's beloved bedtime story. In *Goodnight iPad*, the serenity of the great green room was blasted away. The tussling kittens were gone. Instead of a little mouse, there was a robotic rodent with an antenna emitting sound waves. The snapping fire in the old illustrations appeared in the spoof as pixels on a screen. And where one little rabbit was preparing for sleep in a room that got darker and quieter with each turn of the page, a whole family of rabbits was sprawled around, each oblivious to the other, in a "bright buzzing room" full of technology. The machines in the illustrations with their cables and wires now look outdated, but *Goodnight iPad* was onto something: this is the way we live now.

Rare is the household that does not show the imprint of technology. To the optimistic cultural commentator Virginia Heffernan, the Internet is "the latest and most powerful extension and expression of the work of being human." Maybe: at any rate, in an evolutionary eye-blink, a vast amorphous network has engulfed modern societies. If there were an easy escape from it, most of us would not care to try. The Internet makes possible all sorts of wondrous things. We can search anything, see anything, communicate with anyone. And yet . . .

It is possible to appreciate the digital realm, with Heffernan, as "a massive and collaborative work of realist art," while also harboring unease at the pace and intensity of its advance. Screens have rushed into childhood at the pace of an avalanche, making children more likely to spend their time online than any other place. According to Jean Twenge, a professor of psychology at San Diego State University, the amount of time kids spent online doubled between 2006 and 2016. In 2008, about half of all high-school seniors used social media. The proportion now is upward of 80 percent. And none of it seems to be advancing the cause of human happiness. Since 2012, with the widespread adoption of smartphones and tablets, Twenge and her colleagues have recorded a plunge in young people's emotional well-being.

"We found that teens who spent more time seeing their friends in person, exercising, playing sports, attending religious services, reading or even doing homework were happier," Twenge has written. "However, teens who spent more time on the internet, playing computer games, on social media, texting, using video chat or watching TV were less happy. In other words, every activity that didn't involve a screen was linked to more happiness, and every activity that involved a screen was linked to less happiness."

We adults, meanwhile, seem to be addicted en masse to our smartphones and tablets. By one recent measure, the typical smartphone owner is on his device for three hours a day. Forty percent of us spend much more time, up to seven hours a day. As tech writer Adam Alter notes, this means that on average, and without apparent qualms, we are dedicating *a quarter of our waking lives* to our phones. To Alter, this looks not like enrichment but impoverishment. "Each month, almost one hundred hours [are] lost to checking email, texting, playing games, surfing the web, reading articles, checking bank balances, and so on," he writes. "Over the average lifetime, that amounts to a staggering *eleven years.*"

That eleven-year figure pertains to the minutes we give to our handheld devices. Computers, televisions, and gaming consoles soak up yet more adult time and attention. This matters because the hours we spend on screens, unrelated to work, are hours during which we are not doing other things. And it represents a significant chunk of life in which we are not available to people who are physically present in our lives.

There is no doubt that digital communication makes it possible to stay connected with people we love far away. Unfortunately, the same technology has the practical effect of putting distance even between people who share the same home.

*   *   *

NOT LONG AGO, parents arriving at a day care center in Texas were startled to find the following note taped to the glass front door:

> *You are picking up your child! GET OFF YOUR PHONE!!!! Your child is happy to see you! Are you not happy to see your child?? We have seen children trying to hand their parents their work they completed and the parent is on the phone. We have heard a child say "Mommy, mommy, mommy . . ." and the parent is paying more attention to their phone than their own child. It is appalling. Get off your phone!!*

Popular response to the note was fast, furious, and ran a familiar gamut. On one end was the righteous indignation of those who side with children against their tech-ensorcelled parents. On the other was the righteous indignation of those who side with parents against the finger-wagging busybodies who fail to appreciate that a person cannot hang up on a conference call with the CEO just because he happens to also, at that moment, be collecting a toddler from day care. That the public reaction was both so aggressive and so defensive speaks to the degree to which we remain, as a culture, uneasy about the accommodations we're making. There's no disputing the utility of a smartphone for a busy parent—the thing is a minor miracle—but if we draw back and look at the larger effects, it's an open question whether we are taking the hidden costs into account.

It is becoming clear that heavy digital technology use puts a strain on family relationships. When parents and children are interrupted by technology—when, say, a call or text comes in and the adults' attention gets diverted—children resent it. They may not say so outright, but a 2017 Pennsylvania State University study found that children may begin acting out after average or even low amounts of tech-based interruption, or "technoference." This consists of the disruptions that occur, as the researchers wrote, "during face-to-face conversations, routines such as mealtimes or play, or the perception of an intrusion felt by an individual when another person interacts with digital technology during time together." In the study, researchers tracked 170 families and found that technoference correlated with problematic child behaviors such as whining, hypersensitivity, and flying off the handle.

To children on the receiving end, technoference can feel like personal erasure. Psychologist Catherine Steiner-Adair has seen a steady stream of children in her therapy practice and school consultancy who describe feeling displaced, and confused, too, by their inability to get their distracted parents to pay attention. In her book *The Big Disconnect*, Steiner-Adair quotes a seven-year-old who laments: "A lot of time at home when my parents are home and on their computers, I feel like I'm not there, because they pretend like I'm not there . . . they're like not even talking to me, they just are ignoring me. I feel like, ughhh, sad."

One mother could have been describing the family in *Goodnight iPad* when she described her own to Steiner-Adair: "Our household is eerily silent at night now because everyone is on their own machine, their own little screen. And even after we've put our kids to bed, my husband and I don't spend real time together. We sit at the dining room table, facing each other but staring into our laptops."

This scenario transpires every night in millions of households. Interestingly, researchers were clocking a remarkable degree of matter-of-fact domestic alienation even *before* the mass adoption of tablets and smartphones. Between 2002 and 2005, a team at the University of California documented the interactions of thirty middle-class Los Angeles families from a range of ethnic and cultural backgrounds. The research team videotaped each family over the course of two weekday afternoons and evenings and one weekend. From the outside, these busy, made-for-TV households had everything: decent jobs for both parents, plenty of material possessions, and the temperate loveliness of southern California weather. As a setting for a sit-com, with parents and two or three kids trading wisecracks in a cozy, cluttered family room, any one of these homes would have seemed ideal.

Yet the researchers found a curious emptiness. Members of these normal modern families spent almost no time in each other's company. On average, they spent just 14 percent of their hours at home together. In nearly a third of these households, *parents and kids were never in the same home space at the same time.* As the study authors noted, in the long hours of observation, for these families there was not "a single instance" of full unity or close proximity.

To Susan Pinker, author of *The Village Effect*, these are disturbing realities. "Despite the mountain of evidence showing that family meals and social engagement trump almost everything else in boosting a child's psychological development and health," she told me, "many family members are isolated in their own homes, alone with their personal devices." The world has always had a quotient of dysfunctional, alienated families, she said, but "tech obsessions add a new layer of anomie."

* * *

THERE IS A remedy to this hollowing out of households, a gentle and powerful way to revive personal contact, ward off interruptions, and strengthen relationships. It may seem a small thing, but it is not: we can read together. By setting aside time every day, we can leave the pixelated wilds and rest at least for a little while in a place of unplugged, authentic human connection.

When the writer Michael Sims was young, he had the sensation that he and his mother and the story she was reading were all blending together. "I could feel her voice through my back and side. Her body was part of the story and she made me a part of the story. When my mother paused to take a deep breath, my body rose up a little with hers," he recalled.

I remember the same sensation, reading *Treasure Island* to my family years ago. Molly was pressed against my side; Violet and Phoebe, who were toddlers, sat heavy on my lap; and Paris had draped himself across the back of the sofa, languid as a jaguar. My husband, just home from work, was stretched out on the floor in his suit. It was a golden moment.

None of us knew it at the time, but we were partaking of what clinical psychologist Hilarie Cash calls "a whole bouquet of neurochemicals." Human beings produce this bouquet when we're physically close to people we love, and its effect is to help us stay emotionally and physiologically regulated. We are social animals, after all, not just creatures of social media. Being in affectionate company feels good to us.

Susan Pinker prefers a stronger metaphor than bouquet. She told me: "A tsunami of neurochemical benefits gets unleashed when a parent and child cuddle together over a book. Stress and anxiety downshift, for starters. As soon as the parent puts his or her arms around the child hormones flood their bloodstreams, relaxing them and engendering mutual trust."

This is the chemical explanation for Kate DiCamillo's description of existing together in a patch of warmth and light. The book itself gives off a kind of warmth, like a little campfire, because it intensifies natural feelings of shared purpose. If it happens to be a picture book about love, and loving feelings, and be filled with intentional, affirmative love language—books like Mary Murphy's *Utterly Lovely One*, or *Guess How Much I Love You*, by Sam McBratney, or Amelia Hepworth's *I Love You to the Moon and Back*—a child gets the extra pleasure of hearing tender words spoken in a parent's voice. "The script is right there to start having an emotional connection," in the words of read-aloud advocate Katrina Morse.

The act of reading together secures people to one another, creating order and connection, as if we were quilt squares tacked together with threads made of stories. That's not just another metaphor, as a team of neuroscientists at Princeton has discovered. Even as a reader and listener are enjoying their bouquet of neurochemicals, or being swept away by a tsunami of them, their brain activity is synchronizing, creating literal order and connection in a process known as neural coupling.

"Storyteller and hearer are connecting in a deep way," Geoff Colvin writes of the phenomenon in his book *Humans Are Underrated*. "We not only experience the story; we and the storyteller are having the same experience."

It should not be surprising, then, that the emotional rewards of reading aloud are wildly out of proportion to the effort it takes. When we settle in with a book and a child or two, we get to engage in a "kind of conspiracy," as the Australian writer and illustrator Mem Fox puts it, in which we "bond closely in a secret society associated with the books we have shared."

Incredible to relate, reading books out loud can establish an emotional connection even when the listener is too small, too frail, and too young to know what is happening.

* * *

"WHEN I GOT here, they said I should talk to them, and you can only think of so many things to talk to babies about before you run out of topics, so we started reading to them," Claire Nolan told me when I met her in the neonatal intensive care unit (NICU) at Georgetown University Hospital in Washington, DC. The room was hushed and busy. Doctors, nurses, and visitors moved around in the twilight, speaking in low voices. Monitors beeped and chirped like the tech in *Goodnight iPad*.

Nolan was sitting in a little valley between two large pieces of medical apparatus, with a tiny son, Dale, swaddled in her arms. His twin, Tyrone, was resting in the humid warmth of a nearby incubator, one Lilliputian foot glowing ruby-red from a monitor clipped to it. Tyrone was a little wrinkled and delicate to the point of translucence. He weighed one pound.

The boys' mother said: "We started with little books, but they only last, like, two minutes. And then you end up telling them, 'There's a green frog,' and trying to describe the pictures. It's more fun to read to them. I tried reading them adult books, but they got too adulty—even though they can't understand," she said with a smile, indicating the babies. "I didn't want to be telling them about that stuff." So she and her husband, Jason, started reading *Harry Potter and the Sorcerer's Stone*.

Dale was luckier than his brother from the get-go because he weighed more. His parents could hold him close while they read. Tyrone, who was born a third of his brother's size, was still bristling with tubes and hard to cuddle. He could, however, be touched and caressed by his parents' voices.

"We know that parental voice is important in neurological development," Dr. Mohammed Kabir Abubakar told me as we stood together in the NICU a short distance from the Nolan twins. "There is some assumption, though we don't know for sure, that babies in the womb can hear the voice. How muffled it is, we don't really know. Nobody has been able to put a sensor in there," he said with a chuckle, "and measure the decibels or how much exactly babies hear or how they interpret that sound. Is it just vibration that they feel? Is it the sound that they feel?"

We know that newborn babies are primed to recognize the voices of their parents, in particular their mothers, thanks in part to research at the University of Montreal. Writing about a 2011 experiment there, Susan Pinker notes that "the language circuitry in a newborn's brain comes alive at the sound of its mother's voice. . . . Compared to what happened when a female stranger's voice was played, a brief snatch of the mother's voice provoked a dramatic neural response in the tiny subjects."

* * *

THE EARLIER A child arrives in the world, the greater the odds against him. Important brain development takes place during the final weeks before an infant comes to term at forty weeks' gestation. A baby born at even thirty-six weeks has a heightened chance of developmental and academic delays before the age of seven, according to a 2011 study at Queen's University Belfast. The more parents and doctors can do to encourage brain activation, the better.

As a piece of medical technology, a copy of E. B. White's *Stuart Little* or George Selden's *The Cricket in Times Square* would seem to be out of place among the sophisticated machinery of the NICU. A book is a primitive artifact made of paper, ink, and glue; it's the stuff of kindergarten art projects. Yet in concert with the human voice, this quaint, Gutenberg-era object is a potent tool for stimulating infant brains—and kindling the emotional connections that babies need to develop with the people closest to them.

*   *   *

IN THE SPRING of 2017, Georgetown researchers finished a small study into the effects of reading aloud on twenty babies between the ages of twenty-six and thirty-four weeks. The babies were monitored for ninety minutes, with machines capturing and recording measurements of their physiological state; that is, their heart rates, blood pressure, breathing, and oxygen levels. It was not easy to arrive at a standard method of reading. Some of the babies had been flown into Georgetown from smaller, distant hospitals, and their families couldn't come in often, or at all, to read in person. There were concerns that opening incubator portholes to read to the babies would cause the temperature inside to fluctuate. No one wanted a delicate occupant to get cold.

In the end, the team settled on using recorded narration. Parents taped themselves reading aloud for thirty to forty-five minutes from whatever source material they liked. One mother, a medical student, read passages from a neurology textbook. Some people read children's books or prayer books, and some even read to their babies from the *Wall Street Journal*.

"Inside the incubator it gets really humid, and we wanted something that didn't interfere with any of the equipment," said Dr. Suna Seo, a neonatologist who worked on the study. "We realized we could load the books onto a small iPod, like an iPod Nano, and then have Bluetooth connected with a shower speaker that attaches to the incubator." She laughed. "We have a sound meter that measures the decibels so that they fall within the parameters of the protocol, and then the nurses are able to play the recordings when the parents aren't there.

"They just love listening to the parent's voice," Dr. Seo told me. "We've seen some amazing things." One girl had been born at twenty-five weeks and had suffered complications, including bleeding on the brain. Two weeks later she was stable and snoozing when a nurse pressed a button and her incubator was filled with the sound of her mother's voice. In an instant, the tiny girl had shot to alertness and had begun groping around inside her incubator. The voice of her mother was not a bit of background noise. It was not a matter of indifference. The intonation reached her baby brain and, we can assume, set it sparkling in the sort of dramatic neural response the Montreal researchers had seen.

The doctors and nurses at Georgetown noticed something else: the voice that excited the brain had the paradoxical effect of soothing the body. Babies in the study registered fewer physiological fluctuations during and immediately after the readings. "We can see that during the duration of the reading you have fewer events, of changes in oxygen saturation—it's much more stable, the breathing is much more stable, the heartbeat is much more stable," Dr. Abubakar told me. "And it had a lasting effect for at least an hour after the reading." He knows it lasted for an hour because that's how long the monitors kept running. Had the tabulation been extended, it's possible that the babies would have been seen to enjoy the calming effects even longer.

To appreciate the significance of these preliminary findings, it helps to keep in mind the fragility of the babies. For you or me to have stable oxygen saturation, regular breathing, or a predictable heartbeat is a matter of course. For precarious infants inching their way toward the date at which they ought to have arrived in the world, when all their systems would ordinarily have been at a safe point of development, these can be hard-won achievements. Think what it means, then, for a baby's heart rate and respiration to stabilize when she hears her parents reading out loud. Their voices are a kind of curative.

"Reading promotes better interaction between parent and child. We know for sure that it does," Dr. Abubakar said. "It makes the parent much more actively involved in their child's development. We think if you establish this relationship much earlier, it's much more likely to continue as the baby goes home. And if they keep reading to that child, it's going to promote better child and parent interaction, and that will definitely have an effect on their development, but also on their intellectual abilities."

The story of one Georgetown NICU patient, Sam Green, would seem to bear this out. Twelve years ago, Sam came into the world at twenty-eight weeks weighing just two pounds, four ounces. A photograph of him from the time shows a tiny, almost spectral creature, curled up and stuck full of tubes. Sam's mother, Lori, had read every night to his two older sisters, and she was determined to do the same for him. Day after day, hour after hour, she sat in the NICU with her son's body pressed against her bare skin, in the therapeutic practice known as kangaroo care, and read out loud. There was no way of telling if Sam knew or noticed, but the reading at least gave her a boost. "It helped me bond with him, made me feel useful, and gave our unfortunate circumstance the illusion of normalcy," she told me.

Sam's physical complications took the form of bronchopulmonary dysplasia, a lung disorder that followed him out of the hospital after his NICU "graduation" and kept him attached to a nasal cannula that supplied oxygen until he was eight months old. When he started talking, he had a mild speech impediment. So far, so normal, as far as the bumpy road of preemie childhood is concerned. Yet Sam's brain didn't show any impairment. He never had any cognitive delays.

Could it have been the reading? In those long weeks of dangerous fragility, as he lay in his mother's arms, was her voice awakening his brain in ways that helped to compensate for his precipitate arrival? It's impossible to tell in the case of one child. There's no way to know how a boy's intellect would have developed in the absence of his mother's reading. No test now could show that it made a difference then, nor is there a way to separate the good her reading might have done from the healing analgesic of holding him close. We can't prove a negative, but the research at Georgetown suggests that we can anticipate a positive.

*   *   *

WHEN A NEW baby listens to an adult reading, he may be getting a therapeutic boost, but he's not getting much, shall we say, literary benefit. The hours that the Nolans spent reading J. K. Rowling to their sons in the NICU will be consigned to that inaccessible vault that holds everyone's first experiences. There's no retrieving those files.

But as babies grow into toddlers and then into children and then, incredibly, into adolescents, the picture books and novels they share with their parents and siblings produce a special kind of adhesive. It builds the family, helping to create that secret society Mem Fox talked about, with its common store of words, scenes, and characters.

As members of that society, parents benefit, too. The time we spend reading to our children can feel like a return journey to destinations we visited long ago and never thought we would see again. We may find ourselves soaring, with Sinbad the Sailor, through the skies, lashed to the claw of a giant roc; or resting in the shade of the cork trees with Ferdinand the bull (and noticing Munro Leaf's message of pacifism in a way we didn't when we were little); or perhaps tiptoeing with Bluebeard's bride down a dreadful corridor of locked doors (thinking, Yikes, this would make a terrifying movie).

In Bruce Handy's book, *Wild Things: The Joy of Reading Children's*

Literature as an Adult, he writes, "One of the unexpected joys of parenthood, for me, was reencountering books from my childhood that I had loved and that, much to my relief, I found I still loved.

"Aside from the immediate pleasure of sharing great stories and art with my children," Handy goes on, "these nightly readings gave me a chance to reconnect with books I had loved as a boy and to discover the great wealth of children's literature published in the decades since I had moved on to more 'mature' works."

As the poet William Wordsworth observed: "What we have loved, others will love, and we will teach them how." It's a nice encapsulation of what can happen when parents read with their children. The enchanted hour might be a father's best opportunity to get out his paperback of *Rip Van Winkle* and share his love of Washington Irving. It might be a chance for parents to sell their children on the "real" William Steig, whose writing in *Dominic* and *Abel's Island*, and the peerless *Sylvester and the Magic Pebble*, is so much wiser and funnier than they might guess from the "Shrek" movie franchise inspired by one of his picture books. For a mother with a fondness for nonfiction or science fiction, for nonsense verse or fairy tales, for Christian hagiography or the *Bhagavad Gita*, reading any of these to her children creates invisible threads that will connect them to her, and her to them, and all of them back to the texts in a way that is unique to that family.

When parents and children know the same secret-society books and comic lines, it has a wonderful way of reinforcing a sense of intimacy. I know this from experience. Not long ago I was hiking with my three younger daughters, and idly tapped my leg for a little as I walked along. "Pippi beat time with her false arm," Flora narrated softly. It was a line we all knew from Astrid Lindgren's *Pippi Goes on Board*, and it made us all laugh.

When Paris was home from college recently, ravenous as usual, I had to break the news that dinner wasn't ready. He sighed dramatically: "Things are not very good around here." Out of context, the remark would have sounded rude, but we both knew he was quoting Russell Hoban, from the picture book *A Baby Sister for Frances*, so I knew to supply the next line: "No raisins for the oatmeal . . ."

Every read-aloud family has an anecdote. One woman told me how she'd loved her father's flamboyant accents when he read to her. He would deliver Louise Fatio's *The Happy Lion*, which as anyone can see is set in a pretty French town, with a John Wayne–style drawl. Another father would squeeze himself and his kids into the lower half of a set of bunk beds to approximate the feeling of being "belowdecks" when he was reading *Mutiny on the Bounty*. A dear family friend, a young lady named Beatrice, looked forward to her mother's readings of *Charlie and the Chocolate Factory* because whenever the word *chocolate* appeared in the text, her mother would pop a morsel of Hershey's into her mouth. "She also had a box of Turkish delight for when the white witch offers it to Edmund in *The Lion, the Witch and the Wardrobe*, Beatrice told me. "I loved it, but my brother and sister thought it tasted like soap."

At our house, we've tended to return to particular classics in a rotation of our own peculiarity. Every two or three years for the last twenty or so we've reread the Chronicles of Narnia (except *The Last Battle*, which my children have banned on account of its being too sad) and all the Little House books by Laura Ingalls Wilder. Some years ago I sobbed repeatedly through the Oscar Wilde stories "The Happy Prince" and "The Selfish Giant." We have drifted every few years in a sampan with little Tien Pao behind Japanese lines in Meindert DeJong's *The House of Sixty Fathers*, and rejoiced over the description of "a box of chocolates about a foot square by six inches deep, swathed around with violet ribbons" in Joan Aiken's *The Wolves of Willoughby Chase*. Robert Louis Stevenson's *Treasure Island* is our literal desert island novel, the one book everyone could want if we were shipwrecked (*Kidnapped*, with its confusing Jacobite politics, comes around a bit less often).

We have cycled through *The Secret Garden* and *A Little Princess* by Frances Hodgson Burnett, Rudyard Kipling's *Just So Stories* and "Rikki-Tikki-Tavi," and *My Father's Dragon*, by Ruth Stiles Gannett. Every few years we've dipped into Greek mythology via *D'Aulaires' Book of Greek Myths* and into Homer, in the last few years, through Gillian Cross's superb children's adaptations of *The Iliad* and *The Odyssey*, both illustrated by Neil Packer.

Of course, what thrills in one family may bomb in another. Not long ago, Flora and I decided that we really ought to try J. R. R. Tolkien's *The Fellowship of the Ring*. Having read *The Hobbit* a couple of times over the years, we had only scratched at the topsoil of Middle Earth, and I felt remiss in not having dug deeper. Tolkien had meant a lot to people on my mother's side of the family, I knew. So, spurred by a mingled sense of duty and curiosity, I got us a copy of the first book in the Lord of the Rings trilogy, and off we went.

I am sorry to say that it was hard for us to catch the rhythm of the story. I found it a struggle to bring life to the text. There's one passage, quite early on, in which Tolkien describes the possessions that the old hobbit Bilbo Baggins has bequeathed to his friends and relations after he disappears with a flash of the magic ring. Those six paragraphs seemed to us to go on for six pages. I noticed that Flora was fidgeting. We kept reading, but the enterprise was losing steam when a name in the text gave me a sudden shock. The hobbit Frodo and his companions are trying to reach an important destination. To my amazement, the name was a precious and familiar one: Rivendell.

That was the name my South African grandparents had given to the tiny cottage in Ireland to which they had moved when I was a girl. On the wall of my office today hangs a framed black-and-white aerial photograph of their Rivendell, alone on a scrubby hill facing the Atlantic Ocean; the "last homely house east of the sea," in Tolkien's words.

Somehow my mother, who moved abroad in her late teens, had missed the Middle Earth mania that swept her household. She'd never read Tolkien herself, so she hadn't thought to read him to me. It felt strange and sad to make the connection too late to discuss it with my grandparents. I felt guilty, too; how could I not love this book they had loved, that, indeed, the whole world seems to love? Yet here we were, Flora and I, picking up *The Fellowship of the Ring* with more reluctance every night. Two hundred pages in, we neither of us felt invested in the story. At the same time, we were ashamed to abandon it. How shallow were we, that we could give up on a tale that has delighted generations? Shallow enough, I guess: we bailed.

I tell this story in the context of reading aloud being a means of building relationships—or of missing the chance. Flora and I will always be able to laugh in private about an excess of Misty Mountains. But I cannot shake the feeling that I lost a potential point of connection with my grandparents. I had spent long, happy times with them at Rivendell with no idea that the name meant anything special. Now I'd come to it too late. The spell of the book on my family, at least, was broken. Flora certainly won't be passing it on.

(Still, I am sorry, Gran.)

* * *

IT'S ONE THING to have missed out on a point of connection with grandparents who have passed away. It is another and sadder thing when distance creeps into current relationships. Reading together may have a lovely capacity to fortify emotional bonds, but when family members are separated—by divorce, illness, deployment, incarceration, business travel—it may be impossible for people to sit down together to share a story. Fortunately, there are ways around that. Reading aloud with technology may be less than ideal, given the risk of technoference, but reading aloud *through* technology is an phenomenal backup when the real thing can't happen.

In the spring of 2017, Marine Corps commandant Robert Neller stood before an audience in Washington, DC, and talked about the domestic sacrifices of military families. "We miss sporting events, recitals, graduations," he said of the men and women in uniform. "And while our children miss us with the big stuff, they also miss the everyday stuff: family dinners, homework, kids waking up on the weekend and jumping into the bed, and always the stories. Our kids are the ones that face the anxiety and stress that comes with their parents being deployed. I mean, we're focused on the mission, so we're absorbed in the mission, and that's what your time is, but your children don't know that. They just know that you're not there."

Millions of American children have a parent serving in the armed forces, and a typical deployment runs between four to eighteen months. Between 2001 and 2010, some two million service members served in Iraq or Afghanistan. Just over half were married at the time, and about 44 percent had at least one child. For the people in charge, it was important to know how families were coping with deployments. In 2016, investigators with the Rand Corporation released a report detailing the results of a three-year study that was the first to track children and teens in military families before, during, and after parental deployments. The researchers then compared those youngsters with children whose parents did not deploy.

It is a fact of military life that children face "unique stressors," as the Rand authors put it, in the form of "periodic, extended separation from one of their parents," which can produce "adverse child emotional, behavioral, and academic outcomes." Past research has linked parental deployments to anxiety, depression, and aggression in children, along with attention and school difficulties and conflict within families. Military kids are not alone in this, it should be said, but nonetheless, it's what they're up against.

"Jack, my youngest, who's now seven, had terrible separation anxiety," Alice Kirke told me. Her husband, Kevin, a major in the marines, had deployed to Afghanistan when Jack was eighteen months old but had spent the previous six months at a distant duty station. In other words, Jack was a year old when his father disappeared from his day-to-day life, and it would be another year before he saw his father again in person. Jack did see Kevin on a screen. Under the auspices of a military charity dedicated to keeping kids and parents connected through read-alouds, Kevin recorded himself reading a handful of picture books for his little son, as he had for his daughter in earlier years. Alice made sure that Jack watched the videos while his dad was away.

"But I was really nervous when Kevin came home," she told me. "He had a two-week R&R period. Jack was two, and I was scared about how Jack would react to having Kevin in the house for more than just a weekend."

What happened after dinner on the first night thrilled everyone. Alice had bathed Jack in the bathroom off the master bedroom and fitted him with a diaper. The minute she was done, she said, "he got up and ran out of the bathroom, through the bedroom, past Kevin, who was on the floor stretching, out of the bedroom, down the hallway and into his room." Jack pulled *Curious George* out of his bookcase—it was the last book his father had "read" to him electronically—"then ran back to Kevin, turned around and backed into Kev's lap, sat down, and handed him the book."

Reading long-distance had *worked*. It had kept a connection between father and son. Kevin Kirke was away from his wife and children for the better part of a decade (including three of Jack's first five years). Yet the family has experienced none of the reintegration problems catalogued in the Rand report, such as confusion, alienation, conflict, or depression. Both parents are convinced that it was the reading that protected them.

"Even though Kevin was reading on a screen, Jack wanted to

be with him, physically, to have *Curious George* read," Alice told me. "I always read to Jack physically, so he knew that this is how you must read when you are in the room. Reading became a safe activity, a comforting activity to do together."

Books helped in a different way for the couple's eleven-year-old daughter during the same deployment. A precocious reader, Madison wanted to tackle *The Hunger Games*, by Suzanne Collins. "Kevin was able to read along with her, and there were a couple of chapters of the book that he recorded himself reading, and asking her questions about those things that he read, and furthering that conversation with her. So even though she didn't need him to read aloud because she can read on her own, he still read and was able to engage in those ideas: this is what goes into military decisions and political decisions, and bring it down to her level. The book offered that more profound experience."

The military charity that facilitated the family's interactions, United Through Reading (UTR), was founded in 1989 by Betty Mohlenbrock, a mother whose daughter had not recognized her dad, a naval flight surgeon, when he returned from Vietnam. Mohlenbrock hoped to spare other military families the same pain, using books and reading as a kind of salve. Headquartered in San Diego, UTR has established recording locations on almost all navy ships, at most Marine Corps libraries, and in a wide network of army garrisons. Uniformed men and women can make recordings at seventy-five centers run by the United Service Organization— the military support nonprofit—as well as through roving USO caravans in Iraq and Afghanistan.

Taylor Monaco, UTR's director of communications, told me about a small transition team on the border of Iraq and Syria. "Every time they got new books they would set up a tent and the camera equipment and the marines would shuck off their packs and sit down in this little tent, read a book and be recorded, and those DVDs were sent home via their supply chain," she said. "Even in the middle of nowhere on this border, we were able to support that effort. They could sit down and be parents for five minutes."

The program seems to have had a stunning effect. In 2017, UTR surveyed three thousand participants in the program. Ninety-eight percent of parents reported a decrease in their children's anxiety about deployment; 99 percent said their kids felt more connected with the parent who was away; 97 percent of them said their own stress levels went down because of the reading; and 99 percent of the survey respondents said their children showed greater interest in reading and books.

"It's the culture of shared stories," Monaco told me. "That love of reading gets passed down. Time spent reading together creates a special, magical place. And it's irreplaceable."

*   *   *

THE SAME MECHANISM that helps children and parents in military families also helps to sustain emotional connection for kids and parents who are kept apart by incarceration. Some 2.7 million American children have a parent in prison. Nonprofits active in Delaware, Minnesota, Illinois, California, Kentucky, Vermont, Oklahoma, Texas, New York, New Hampshire, and elsewhere (including in the UK) are making it possible for thousands of inmates to record themselves reading children's books. In many cases, volunteers can send children not only the recording of a mother or father reading but also copies of the books.

The great-grandmother of the movement, in a manner of speaking, is Aunt Mary's Storybook, a Christian prison ministry founded in 1993 at a women's correctional facility in Cook County, Illinois. The group operates in sixteen jails and prisons in Illinois, and inspired the creation of similar programs in Kentucky and Texas.

"From the womb, the children know that voice, and this is just keeping them together in the best way we can until they can be together," said Stuie Brown, a grandmother who's been part of the outreach in Kentucky for two decades.

"These young mothers are incarcerated and something bad has happened," she told me. "Now they are on video, they take such time to wash their hair and look as pretty as possible. They so desire for their children to see them."

In a protocol that is typical of prison reading programs, Brown and her fellow volunteers are allowed to enter the Kentucky Correctional Institute for Women, outside Louisville, a few times every year. The volunteers have to undergo extensive training and vetting, and they are frisked and patted down before being admitted. The rules of engagement are strict, too. The volunteers and the inmates are allowed no more contact than a brief exchange at the start and finish of the recording session.

During each day-long visit, the volunteers record the readings of about twenty-five mothers. The women are allowed to introduce each book with a short remark. "This is Mommy. I chose this book and I love you and I miss you and I'll see you soon," Brown said, by way of example. "They're allowed to read picture books all the way through, but with novels the moms usually just read the first chapter and encourage their children to keep going from there."

"It's hard for me to explain, but I think something huge is going on," Brown said. "It's not an easy thing. I can't know these women or their children, I'm just a nobody. But it has increased my faith. I feel like my little part is that I'm keeping that mother and child connected.

"I get to carry books and sit in a chair and listen," she said softly, "and cry when I go home."

*   *   *

ESTABLISHING A STRONG relationship with a parent or a child or a spouse (or with anyone else, for that matter), is not a one-time event. Tempting though it may be to believe otherwise, psychology teaches that an emotional bond isn't so much an end point as a state of being that we have the power to neglect or enhance.

Sitting together, reading together, focusing on the same story; these are intimate acts that bind us closer to one another. If a relationship is troubled, or if illness or disability or adolescence makes talking awkward—if conversation feels too much like pressure—a shared book can lift everyone up and out of the situation, spanning the difficulty like a bridge.

*   *   *

DANICA AND ERIC Rommely did not find their way to that bridge until their son Gabe was a teenager. At seventeen months, Gabe had been diagnosed with severe autism. As the years ticked by, he didn't talk, and he didn't seem to understand what people were saying around him.

"We read him board books when he was young," Danica told me, "but he would never sit still, and at some point, when he was a toddler, we thought, screw this, he's not listening."

Gabe's main interest was watching TV shows for little kids. He would watch episodes of *Sesame Street*, *Barney*, and *Blues Clues* over and over, sometimes on several devices at a time. When his parents or caregivers took him to the library, even after adolescence he went straight for the baby books.

"So that's where I thought we were," his mother said. "That was his level."

Except, it turns out, it wasn't.

*   *   *

HELLO . . . MEGHAN . . . *how are you . . . on this fine afternoon?*

The voice belonged to a young therapist named Najla. The words, typed with one finger on a wireless keyboard, were Gabe's.

He was being funny: it was not a fine afternoon. It was beastly late January, and outside the Rommelys' house freezing rain was sluicing down, gurgling in the gutters and spattering the leaves.

Everything had changed for the family eighteen months earlier when a therapist introduced Gabe to a system called the Rapid Prompting Method, the mechanism that he and Najla were using when I met them. For the first time in Gabe's life, he could express himself in full sentences. Until that point, for fourteen years, he had not been able to show that he *had* been listening, he *did* have feelings, and though he could not speak, he did have a voice—a voice that was witty, intelligent, and highly self-aware.

As you may imagine, Gabe's parents were at once stricken with remorse and suffused with gratitude. They now had a chance to get to know their son in a fresh and wonderful way, and he, at last, could express himself despite being trapped in what he has described as the body of "a drunken toddler."

Danica and Eric began reading to Gabe again. They would sit with him on the sofa, and now, though he might rock or fidget, they knew it didn't mean he wasn't paying attention ("Even though I don't seem to be listening, I am," he said). A couple of times a week, his therapists would read magazine and newspaper articles out loud to him. His parents chose young adult literature. By the time I met them in the kitchen of their suburban home, they'd read the first book in the Harry Potter series, they'd tried *The Hunger Games* ("but that kind of fell apart," Danica said), had devoured Lois Lowry's *The Giver*, and were about to embark on *The Absolutely True Diary of a Part-Time Indian* by Sherman Alexie.

Using the keyboard, with Najla speaking his words at the pace he typed, Gabe explained: *I am addicted . . . to . . . screens . . . but there's nothing . . . I'd love more . . . than to have . . . someone . . . read to me . . . all day long . . . instead.*

It is hard to do justice in writing to the poignant pace of our conversation. It was obvious that Gabe had a quick mind, but he could express himself only at the speed of a one finger hitting one key at a time.

I wish I could read . . . on my own . . . but . . . my body doesn't cooperate.

I asked how Gabe felt, on the inside, when his mother read to him.

It takes me to another . . . place . . . in which I'm . . . completely . . . normal.

Gabe was still typing.

Of course, I love . . . being close to her . . . I experience the world . . . through movies and books . . . to quote a friend.

Najla explained that one of Gabe's friends, who is also autistic, had used a similar phrase.

It was striking to consider that the same technology that has so many of us in its thrall—the screens and keyboards that dominate our waking hours—has freed a personality that had been trapped and silenced. What memoirist Judith Newman calls "the kindness of machines" has for Gabe given voice to an intellect expanded and nourished by reading aloud.

"Do you have anything to say to parents or families with kids who don't seem to be paying attention?" I asked him. "Should they keep reading to them anyway?"

Gabe typed.

A . . . million times . . . yes. We are always listening.

*   *   *

READING TOGETHER CAN do so much important work. It can create connection out of alienation and distance. It can act as a catwalk over the turbulent waters of toddlerhood, and do the same years later in the storms of early adolescence. I'm convinced that reading aloud kept my children closer to me, and to each other, than if we had gone without it. The reading was especially helpful when Paris, my son, was in middle school, that grim epoch when a garrulous young man turns monosyllabic. We didn't have much in common in those years. He was no longer interested in joining his sisters for story time, but he did want to keep reading with me, provided it was just the two of us. So for a year or so, after Molly had gone to do her homework and I'd tucked the younger girls into bed, Paris and I would meet in my office at home, amid the buttes and canyons of books stacked everywhere.

Sitting side by side on an ancient sofa in a pool of lamplight with the night sky black against the window, we read Conrad Richter's heartbreaking novel *The Light in the Forest*, which tells of a white boy raised by the Lenni Lenape who is forced, under a treaty in 1765, to rejoin the settler society he hates. We also read *The Underneath*, by Kathi Appelt, a novel that coils and seethes and builds to a moment of shocking violence and even more shocking mercy.

A lot was happening in the stories, but I now understand how much must have been happening to us, too. We were awash in neurochemical benefits. Our brains were in neural alignment. We were adding to our shared store of references, characters, and plot twists. And to this day, Paris and I are both aware of those evenings as if they were a solid thing. We built something real between us that rested on a foundation of the years of reading that had begun when he and his sisters were little.

CHAPTER 4

TURBOCHARGING
CHILD DEVELOPMENT WITH
PICTURE BOOKS

Here's a little baby
One, two, three
Stands in his crib
What does he see?

—Janet and Allan Ahlberg, *Peek-a-Boo!*

On the opening page of the Alhbergs' classic book, a tiny boy in blue pajamas stands upright, hanging on to the top of his crib with one hand. Mouthing the white wooden railing, he reaches with his other hand toward his father, fast asleep and just visible through a circular cutout in the facing page.

"Peek-a-boo!"

Turn the page of the book, and the aperture falls to the left, encircling the baby and revealing what the baby sees: "His father sleeping / In the big brass bed / And his mother too / With a hair-net on her head."

First published in the UK in 1981 as *Peepo!*, this cozy rhyming tale traces the events of a baby's ordinary, exciting day: waking up before the rest of the family, eating breakfast in a high chair, being taken for a walk, sitting with a toy on a blanket in the park, and being wheeled home for supper, bath, and cuddles before bed. For adults, the book is an exercise in nostalgia, either real or imagined. The family in the book is British, and we can tell that their shabby, cramped row house has limited plumbing. From a portrait of Winston Churchill on the living room wall and from the uniform the father is wearing in the final picture, we know that the story takes place during wartime. It's clear that after the man kisses his son goodnight, he's leaving the house and joining his military unit.

Simple in appearance, this Ahlberg favorite is a triumph of complexity. Its delicate, detailed illustrations are full of unspoken prompts for questions and quizzes. There's depth and texture in the layering of generational perspectives. The verse bounces along, effortlessly imparting meter and rhyme to the child listening. And then there's the captivating meta-concept of peek-a-boo itself, a game that plays on a child's developing understanding of the permanence of objects. For a brief and happy period, babies think you really have vanished when you hide behind your hands and chortle with laughter and surprise when you pop back into view. The pleasure of the game lingers well after children have grown out of genuine bewilderment, which gives *Peek-a-Boo!* an element of nostalgia for toddlers, too.

As the book reminds us, it doesn't take a lot to fascinate a baby. The little boy in *Peek-a-Boo!* notices all sorts of things that older, more jaded observers might disregard: pigeons on the wing, a passing dog, "the tassels blowing on his grandma's shawl." It may be wartime, but this baby has pretty much everything he needs to thrive. He is surrounded by affectionate family members who play and talk with him. The house may be cluttered, but it's a place of routine, that unsung virtue of normal life that gives comforting boundaries to the vastness and caprice of the world. Yet for all that, something is missing.

Can you guess what it is? Correct. In the *Peek-a-Boo!* household, we never see anyone reading to the baby; not his grandma, not his mother or father, not his two attentive, competitive sisters. The child on his mother's lap who's looking at the pictures with her and hearing the text read in her voice has, in this respect, a huge advantage over his fictional counterpart.

*   *   *

READING ALOUD IS good for people of every age, but its effects are perhaps nowhere more keenly felt than in infancy and early childhood. There are good reasons for this, not least the galloping pace of brain growth in a child's first three years of life. Reading to children during this period gives them more of exactly what they need: more loving adult attention, more language, more opportunities to experience mutual engagement and empathy. Picture books enhance the time parents and children spend together. It's like adding an extra shot of espresso to a café latte: one cup, extra zing.

And then there is the role that stories play in what the novelist Shirley Jackson called "the nightly miracle." Bedtime is manifestly sweeter, happier, and nicer when the electronic devices are put away, the books come out, and everyone settles down. A busy day of transactions—spooning cereal (and eating it), washing faces (and being washed), changing diapers (and being changed)—yields to a quiet time of mutual encounter.

An infant won't know much about any of it, of course. Like the babies in the NICU, he won't remember hearing about Pat the Bunny or a very hungry caterpillar. He won't have any trace of a wisp of a recollection of sitting on anyone's lap and reading *Peek-a-Boo!* or any other book. Yet long before a baby is old enough to interact, before the first reciprocated smile, before he can control his head, or sit up, or muster the motor skills to "find the mouse" in an illustration, he is drinking in sounds, responding to affection, and, through his brand-new eyes, learning to distinguish one object from another and to see patterns in the world. Hogwarts and the Hundred-Acre Wood are destinations years in the future.

But almost from the moment a baby arrives, he's paying attention. And why not? He has everything to learn.

* * *

IF YOU HAVE ever studied a foreign language—in particular a *very* foreign one, with letters and sentence structure unlike your own—you may have had the thrilling experience of hearing distinct words emerging from what had once been a fog of indistinguishable syllables. For me, this happened with Japanese, a means of communication so simple that every toddler in Japan can speak it; whereas I, studying the language for two years starting at the age of twenty-nine, was only ever able to muster anodyne and elementary remarks.

Spoken Japanese at first sounded to me like one spectacular run-on sentence. I couldn't distinguish the end of one word from the beginning of the next, let alone tell which were nouns and which were verbs. But as I practiced and listened, small fragments began to sharpen and catch. Aha, I'd think, that means "river" or "fish" and, wait, that sound isn't a word at all, it's a formalized pause, like "um" or "ah."

The more familiar each of those fragments became, the more capable I was of distinguishing other, different bits. After a time, I could make out distinct chunks of grammar, though I remained unsteady with vocabulary. Had I continued adding to my collection of nouns and verbs and adjectives and adverbs and exclamations and, when I was adept with those, picked up colloquialisms and idioms and metaphors, well—that would have been *subarashii* (wonderful!) and *insho-teki* (impressive, superb, magnificent!). Alas, I didn't, so it wasn't. But the experience did give me a taste of what all children undergo in the early years, apart from those who are born with hearing impairments or raised in cruel isolation.

In the beginning, there's the muffled cacophony of the outside world, a mother's heartbeat and her reverberating voice. (At least, that is the assumption, as Dr. Abubakar said.) What follows is an extraordinary, seamless process as the amorphous surrounding sounds of a native language separate into syllables, which at some point in the future become discrete words.

"Language comes at us at an incredibly fast pace, and we need to be able to group things together very quickly, otherwise it's just impossible to understand," said Morten Christiansen, who runs the Cognitive Neuroscience Lab at Cornell University. Learning a first language is a complicated developmental exercise that involves isolating sounds and identifying them with meaning, retaining those first conclusions while isolating new sounds and assigning to them yet more meaning, at the same time adding meaning to the existing stock of words. This process, which neurologists call "mapping," proceeds at two speeds. First comes "fast mapping," when a child will form a hazy hypothesis of the meaning of a new word. During "extended mapping," also known as slow mapping, the child incorporates the word into his memory in a more gradual way while refining his understanding of its meaning.

These calculations and adjustments take place at lightning speed all the time a small child is awake and hearing words spoken aloud. "We do know that the amount of exposure to language that kids have really matters," Christiansen told me. The more speech that a child hears, the greater and earlier his chance of mastering it.

This needs to be understood in a particular way. Ambient talking seems to do little or nothing for babies and toddlers. If two adults stand around talking to one another, the baby in the bassinet is likely to tune them out. What helps babies most is having people speak and read *with* them, in a responsive way. As one academic observed, "If hearing language was all that mattered, children could be set in front of a television or radio to learn their native tongue."

They can't. Babies don't learn from machines—at least not yet. What millennia of human experience and innumerable modern studies do show is that they learn from *us*. They need us to pay attention to them, to talk and play and read with them.

*   *   *

NEWBORN BABIES AND newly hatched songbirds do not look alike. They have something in common, though, apart from being new in the world. They both require instruction. Without a teacher of his own species, a child will not learn language, and a songbird will not acquire the distinctive trills of its kind. (In most songbird species, teacher and student also happen to be father and son. Attracting a mate is a signal purpose of warbling. So when an elder zebra finch or white-crowned sparrow teaches a younger one how to sing, he's literally demonstrating "How I met your mother.")

Baby birds raised in isolation can't make up for lessons they missed afterward. "If a bird doesn't hear the tutor, it will sing," acceded a researcher at Penn State, "but its song will be nothing like the song of an adult bird. It will be poorly structured and lack the wealth of acoustic structure."

Something similar happens with human fledglings. To learn the "song" of their species, they need to be brought into the give-and-take of speech. "There's a lot of language learning that's social in nature. One of the first things that we learn as children is, actually, the social part of it," Christiansen said. "Very early on, we learn the normal, it's-your-turn-it's-my-turn way of talking. So when their little kids are sort of oohing, parents will ooh back to them, in patterns that are reminiscent of how we communicate when we are talking."

Infants also learn from seeing expressive human faces, with their thousand-and-one unspoken inflections. What matters for the child's learning is contingency and responsiveness. Children raised without much of this will learn to communicate, of course, if not as well. A young mind is a thirsty thing. Like a tree growing in Brooklyn, it will draw what sustenance it can, even from chalky soil and a shady setting.

That being said, in recent years we've gained a more profound understanding of what babies and children need to thrive from the tragic example of those who were denied it. The 1989 fall of the Communist regime in Romania laid bare that government's practice of warehousing orphans from infancy onward. Tens of thousands of children were found living in bleak institutions, deprived of affection, stimulation, and personal connection. Babies lay in cribs on their backs, staring up at blank ceilings. Toddlers sat alone in iron cots without toys or books, sometimes tied to the rails so that they would not climb out. Feedings were conducted in silence. During outdoor breaks, adult caregivers spoke to one another, ignoring the children, who milled around without direction or purpose.

In subsequent examination, the young inmates were found to be severely compromised, often having low IQ and suffering from a constellation of psychological, neurological, and biological impairments. The Romanian orphans were like songbirds raised in isolation. They could make noises, but they couldn't sing.

"When the child vocalizes and the caregiver responds, the child becomes a partner in the language experience. In an institutional setting, harried caregivers who are responsible for changing a room full of children cannot make time to respond to the communication attempts of each child. As a result, the infant eventually stops trying to communicate," write Charles A. Nelson, Nathan A. Fox, and Charles H. Zeanah in *Romania's Abandoned Children*, their powerful account of a landmark twelve-year study of these children and efforts to rehabilitate them.

The orphanages of Bucharest offer a distressing illustration of the damage that can be done to children if grown-ups don't engage with them. It's a reminder that babies don't come into the world destined to express and master themselves. They need the people around them to kindle their little brains and show them the way.

*   *   *

CAN TECHNOLOGY HELP with this?

Yes—but also no.

Yes, parents can buy apps and games that promise to stimulate their infants' brains. Lively programs and child-friendly devices purport to be able to teach babies and small children their colors, their math facts, and the principles of sustainable ecology. According to *Parents* magazine, which you'd think would know, "Fun and kid-friendly iPhone applications keep your tot busy and learning on the go." Parents.com promises that "your child's education doesn't have to stop after school," because "games made for your iPhone, iPad, and Android will keep your youngster's mind active outside the classroom."

These claims are seductive, but there's one thing the purveyors of electronic toys and child-oriented tech would rather not say. In truth, their products are inferior. They are no match for *us*. There is no genius in Silicon Valley who has yet devised a machine half as effective for teaching and nurturing the young mind as a flawed, fallible, physically present human being. A real mother who talks and reads picture books beats a Little Mommy Talk with Me Repeating Doll™ every time.

Numerous experiments bear this out. In 2010, a team at the University of Virginia investigated the effect of a bestselling DVD that promised to teach vocabulary to infants. For the study, the team enlisted ninety-six families with babies and toddlers between the ages of twelve and eighteen months, and divided them into three groups. In the first cohort, parents watched the DVD with their children. In the second, the babies watched by themselves. Participants in the third group did not view the DVD, but the parents were asked to introduce the target vocabulary words during normal conversation with their children.

One month later, the researchers tested all ninety-six babies, and discovered that the video had no pedagogical value whatsoever. It didn't matter if the child had watched the DVD alone or with a parent. The words simply did not travel from the pixelated screen to the infant mind. Interestingly, the babies who didn't watch the DVD, but instead heard the words spoken by their parents, *were* able to pick them up.

Another study that same year at Northwestern University, also involving ninety-six babies, tested the efficacy of videos with purported educational value and came to similar conclusions: there was no evidence that babies learned from the screens. A 2007 study concluded that infants and young children learned six to eight *fewer* new vocabulary words for each hour of baby DVDs they watched than babies who saw no videos at all.

According to Catherine Tamis-LeMonda, a professor of applied psychology at New York University, "One of the reasons kids don't learn from media, from technology, is that there's not contingency," the spontaneous, fluid adaptation to what a child seems to be understanding or failing to understand. As director of the university's Center for Research on Culture, Development and Education, Tamis-LeMonda spends a lot of time observing small children, and like many in her field, she worries that in our culture's infatuation with technology, we're not yet getting the balance right.

*   *   *

WE KNOW, IN adult life, that when the screen comes on, the amount of spoken give-and-take tends to dwindle. It's always a bad sign for general conversation when someone pulls out his phone to show everyone a funny video or to read an outrageous tweet. In the fumbling moment between the thought—*Oh, I have to show these guys*—and the electronic summoning—*Wait, it's loading, just a sec*—there's a stalling of momentum.

We know that technoference is a real issue for many children. It can be distressing to lose a parent's attention to electronic interruption, and some kids misbehave because of it. For there to be an interruption, of course, two people have to be engaged with one another in the first place. But what if that engagement is happening less often, or not at all? Is technology smothering some parent-child interactions before they even have a chance to begin?

For a glimpse of a child's perspective, in 2015 researchers at the University of Arizona, Flagstaff, evaluated the ways that families behaved with different kinds of toys. They wanted to see which toys fostered the liveliest chatter between parents and young toddlers. The researchers made in-home audio recordings of twenty-six pairs of parents and children between the ages of ten and sixteen months. Each family was given three sets of playthings: electronic toys (a baby laptop, a baby cellphone, and a talking farm), traditional toys (a chunky wooden puzzle, a shape-sorter, and rubber blocks with pictures), and five board books featuring farm animals, shapes, and colors.

If you've ever planned to "just check email for a minute" and realized with a jolt that you've been disengaged from your surroundings for an hour, you may be able to guess what the Arizona researchers observed. With the electronic toys, the amount of parental chatting and infant vocalizing plummeted. The machines made the noises, the people were silent. There was a bit more talking when the children played with the traditional toys. But the best object of all for eliciting back-and-forth exchanges was a picture book. Reading a board book with a baby turns out to be far more effective than both traditional toys and electronic entertainment in creating an environment rich with words and language.

"These results provide a basis for discouraging the purchase of electronic toys that are promoted as educational," the researchers wrote, and "add to the large body of evidence supporting the potential benefits of book reading with very young children."

One reason baby books are so helpful in this regard might seem to be so obvious as to scarce merit mentioning: books contain words. (Well, of course they do!) But there is more to it than that. Board books in general do not have a great number of words. Often there's no more than one word to a page, or no words at all, just pictures. It's in the *interaction* with the book that the word-magic happens. To return to the example of *Peek-a-Boo!*, the child on the lap will hear the language of the printed text, read by an adult, but also the talking that results from the sight of the pictures. Coming to an illustration of the baby in his high chair, a parent might naturally make a few remarks about the scene: "What's that baby doing? Is he eating his breakfast? Look, he has a spoon in his mouth." This sort of chatter is a huge help to babies as they begin to pick up the rhythms of vernacular speech, which, as Dante observed, is everyone's first language.

To understand why in-person conversation is so constructive—and why screens fail at the task—it may be helpful to consider what happens when babies are taught not a native language, which is the air around them, but a foreign language. In 2003 a team of clinicians at the University of Washington's Institute for Learning and Brain Sciences found that babies from English-speaking families were able to learn words in Mandarin Chinese if they were taught by live, interactive experimenters. These babies were indifferent to and unable to learn from teachers who appeared on screens and who were unresponsive—in other words, not contingent. Twelve years later, the researchers discovered something else. They found for the first time a demonstrated link between a baby's acquisition of the sounds of a foreign language and *the direction of the child's gaze*.

Babies tend to be around two months old when they begin to make eye contact. (It is a thrilling moment for parents: We are *seen!*) By six months, about half of babies will engage in gaze shifting; that is, they can make eye contact and then follow the other person's gaze to take in whatever he or she is looking at. This is one of a baby's earliest social behaviors, and by twelve months almost all babies will do it. (Autistic babies may follow their own time lines in this regard.)

In the study, seventeen infants from English-speaking house-holds, all of them nine and a half months old, were given a dozen Spanish lessons over the course of a month. During each twenty-five-minute session, the tutors introduced toys, read aloud, and chatted in a cheery, interactive way, always in Spanish. The babies wore electroencephalography (EEG) headbands fitted with sensors snugly around their heads. The headgear allowed the researchers to measure the babies' brain activity, while, from the outside, during the first and final sessions, they counted how often the infants shifted their gaze between their interlocutor and the object the adult was showing them.

"How well infants are able to shift their gaze coordinates with their language development and predicts how large their vocabulary will be when they're older," one of the study's coauthors, Rechele Brooks, later told an interviewer.

The brain has a pattern of recognition when they notice, "Oh, you said something new?" Babies in the study were better able to recognize the Spanish sounds. And the babies who have been shifting their gaze more often during play with the actual people were better able to recognize the different speech sounds when we tested their brain reaction. In other words, they had learned the new sounds of the new language, and we were able to recognize it by reading their brain waves during the study.

Watching video excerpts, I was struck by the tenderness of the experiment. It is one thing to talk of studies and findings and the cold statistical implications of one intervention or another. It's quite another to see a small girl sitting on her bottom in that sturdy, pudding-like way babies have, tipping her head in curiosity as a smiling clinician, also seated on the floor, shows her a yellow rubber duck, and then a pretend slice of bread. The girl takes the bread and turns it in her soft little palms, wondering at it. The experimenter, a young woman with a ponytail, leans forward as she chats in Spanish. She taps the bread, as if pointing out its artificiality. She meets the girl's inquiring gaze when the child waves the toy above her head.

"Pan," the young woman confirms. "Pan de jugar." The soft sound of the phrase in Spanish is nothing like the sharper English pronunciation—"toy bread," or "pretend bread"—and having heard it in such a gentle, interactive setting, it makes sense that the girl would indeed remember the sounds when it came time for her vocab test.

Something else is happening in that scene. In exploring an object together, communicating through the voice and through eye contact, the young woman and the child have embarked on a period of what is known as "joint attention." This is a phenomenon that for children has a remarkable tempering power.

*   *   *

IN 1934, A young psychologist died of tuberculosis in Moscow. His name was Lev Vygotsky, and he left this world believing that in his work he had glimpsed, like Moses, a promised land he would never enter. For many decades his thinking had little influence in the West, but since the 1980s, after his writing was translated, his fingerprints have been all over our understanding of child development. Vygotsky believed that play is for children a crucial mechanism for self-discovery. He also believed, signal to our purposes here, that language is a vital tool for a child learning to regulate his emotions and behaviors, and to establish relationships with others. The more adept a child becomes with words, the sooner he can handle himself.

We don't think of babies and toddlers as creatures who can self-regulate. Popular stereotype suggests the opposite: picture a cartoon baby screaming for attention, rays of outrage shooting off his plump Looney Tunes body; or a toddler meltdown, a scribbled hurricane of inchoate rage and frustration. Even if real children seldom resemble the monsters of caricature, there's an element of truth here. As babies and young children grow, they do need to learn to govern their emotions and impulses, to get along with others, and to focus and pay attention. These and other executive-function skills, such as mental flexibility and the capacity to retain information, as well as persistence—grit, or stick-to-it-iveness— are not merely helpful a little later, when children get to school, but important skills for navigating the long business of life.

One of the best ways to help small children maximize these capacities is to read picture books with them, early and often. As one academic team has noted, "Children benefit when they and their parent establish a positive pattern of relating while reading," not least because they "learn to naturally regulate their attention when they are focusing on a task they find interesting in a context that is nurturing, warm, and responsive." Storybooks are less like medicine, in this context, than like vitamin supplements that fortify. (Fast-paced TV shows, meanwhile, have been shown significantly to *impair* executive function in young children after as little as nine minutes.)

We can get a glimpse of the effect of picture books on children's emotional regulation through work done at twenty-two Head Start centers in central Pennsylvania a decade ago. Teachers incorporated a read-aloud scheme that included teaching the alphabet and asking children lots of questions. In the stories, children encountered characters in exasperating situations who were able to model desirable behaviors. For instance, a bright green turtle named Twiggle showed the children how, when he was angry, he would retreat into his shell, take three deep breaths, describe what had upset him, and then articulate his feelings about it. Low-income children enrolled in the centers that used this protocol were shown to make significant improvements in the development of executive function.

"The lesson teaches them to take a time out from their emotions, to avoid acting impulsively," Dr. Karen Bierman, who created the protocol with her colleagues at Pennsylvania State University, said at the time. "Stating what's bothering them, and how they feel, is the basis for self-control and problem solving in stressful social situations."

In 2011, David Dickinson of Vanderbilt University and three colleagues wrote a paper, "How Reading Books Fosters Language Development Around the World." Mustering a cavalcade of evidence, the authors observed, as Vygotsky envisioned, that language "seems to make it easier for children to regulate their own thoughts, feelings, and actions or abilities that are essential to social development and school success."

"The acquisition of expressive language [is] related to less aggression," the authors declare, citing a study that found, in a demonstration of the obverse, that "expressive and receptive language *deficits* in kindergarten predicted later conduct problems." Moreover, children who sustain longer periods of joint attention at eighteen months tend to possess stronger productive vocabularies at the age of two.

"Consider all the ways in which storybooks conspire to help children maintain their attention," the paper continues. "Children's books often use bold colors and strong contrasts and typically depict objects and animals that appeal to young children. The page of the book provides a clear focus for attention, and unlike moveable toys such as balls and trucks, books are held and remain relatively stationary. An attentive adult can easily notice what a child is attending to and build on it with commentary. In turn, children are able to draw an adult's attention to interesting pictures using a broad range of cues including gestures, sounds, and words. Thus, attention can be managed by the child as well as the adult."

When my daughter Phoebe said her first word, it happened in just the way that Dickinson's paper describes. She and her brother and sisters were piled in together as I read *Peek-a-Boo!* Studying the illustrations, Phoebe drew my attention with "gestures, sounds, and words." To be precise, she pulled her fingers from her mouth, poked at a picture, and said: "Goh-goh."

For a second there was stunned silence. Then we all fell about. "Yes! That's *right*, Pheebs! That's a *dog*!"

"Phoebe said her first word!"

"Goh-goh!"

We still use the word to this day.

*   *   *

THERE IS A downside to the phenomenon of joint attention, and of babies learning to look where we're looking. If we are glued to our electronic devices, that's what will draw their gaze, too. And what they see when they look at our faces while we are online may not be what we want them to see. In her psychology practice, Catherine Steiner-Adair met a young mother who, despite her best intentions, could not seem to refrain from getting out her tablet whenever her six-month-old son seemed to be occupied and content. "He's just lying here and playing," the young woman said,

so I'm on the iPad and suddenly he stops playing and he is look-ing at me! I mean so many times—that happens 90% of the time—and I don't know at what point he stopped playing and started looking at me. It breaks my heart because I don't know how long he has been staring at me. I mean, what is he thinking? I feel so guilty that I'm not present with him and he knows it. It's one thing if I'm unloading the dishwasher and talking to him. That doesn't require brainpower, but e-mail does. It's impossible to be really doing both. I know he knows I am completely dis-engaged, you can just see it in his eyes. So what does that mean to him [that] we are both in the same room together and I'm not being present with him?

We know, Steiner-Adair writes, that "babies are often distressed when they look to their parent for a reassuring connection and discover the parent is distracted or uninterested. Studies show that they are especially perturbed by a mother's 'flat' or emotionless expression, something we might once have associated with a depressive caregiver but which now is eerily similar to the expressionless face we adopt when we stare down to text, stare away as we talk on our phones, or stare into a screen as we go online."

As parents in the distracted digital age, we're in an awkward spot. Like the fox in William Steig's picture book *The Amazing Bone*, we may be justified in feeling a little defensive: "Why should I be ashamed? I can't help being the way I am. I didn't make the world."

That makes sense to Dr. Perri Klass, who teaches journalism at New York University, practices pediatrics at Bellevue Hospital, and, as national medical director for the charity Reach Out and Read, helped to write the American Academy of Pediatrics report that encouraged pediatricians to recommend reading aloud. She's sympathetic to the tech predicament parents are in and hopeful that we can find an equilibrium. "All very successful technologies end up being introduced as uncontrolled experiments," she told me,

> and they change our lives and we look back and there's always a certain amount of panic about them.
>
> These technological changes, everybody walking around with the phones, have potential to push people in all kinds of directions but I don't believe it's evil. The dangers for children between birth and the age of three is that all learning happens in a social context. All learning happens through relationships, and the younger you are the truer this is.
>
> The biggest worry, I think, with the technology is that it makes it easier and easier with younger and younger children to replace that interaction time, which is absolutely critical for

all the different kinds of child development, for language, for social-emotional, the development of empathy, and learning how to read people's emotions and faces, and the development of theory of mind—all the different domains come together.

Theory of mind is the understanding that other people have thoughts, feelings, and motives just as we do, but that their thoughts, feelings, and motives may not be the same as our own. Children don't start out with a grasp of this; it's something they learn. Most will progress in a natural way from the sweet obliviousness of believing themselves the center of the galaxy (although, let's be honest, we all know adults who haven't gotten past this yet) to the gradual understanding, at about the age of two, that other people also have wants and needs. This perception deepens and becomes more sophisticated over time, and by the age of five, children tend to be able to understand that their actions may cause others to experience emotions. Theory of mind, empathy, being able to pick up on facial and tonal cues—all these qualities help children to become socially competent.

The personalities and conflicts that children encounter in storybooks can intensify their emotional awareness with amazing rapidity. A 2015 program in the north of England sent trained readers into nurseries that cared for two-year-olds in impoverished areas in and around the city of Liverpool. Over the course of just fifteen weeks, teachers and project workers involved in the scheme reported improvements in the children's language abilities and increased responsiveness to books and storytelling. They also saw, among the parents of these kids, greater enthusiasm and confidence about shared reading.

In the toddlers, contact with the characters in picture books seemed to stir new depths of empathy. As one observer reported: "I was reading *Solomon Crocodile* [by Catherine Rayner] and as the story goes, Solomon annoys all the animals in the river so that they shout 'Go away!' at him. When I got to the page with the hippo and his wide-open mouth, as he roars 'GO AWAY' at Solomon, two-year-old Finn made a pushing action with his hands towards the book and shouted, 'Go away, go away!' at the hippo." Little Finn had taken Solomon Crocodile's side and, in a surge of fellow feeling, wanted to defend him against the rude and shouting hippo.

As Dilys Evans writes in *Show and Tell*, a book about illustration for young readers, picture books "are often the first place children discover poetry and art, honor and loyalty, right and wrong, sadness and hope." And it's true: a child sitting on a lap at home or in the friendly security of circle time at school or at the library has a chance to witness the emotions of others and experiment with his own without consequence. He can try out big ideas, find consolation for his secret worries, and risk a glimpse at what scares him.

I had a dramatic demonstration of this with my daughter Flora when she was four. She was terrified by Donna Diamond's wordless picture book *The Shadow*, yet at the same time found it mesmerizing. Whenever she brought the book to me, she would hold it away from her body with her fingertips, as if it were a great flat spider.

"Why do you want me to read it, if you're not even going to look at the pictures?" I asked her as we settled on the sofa together. Flora's response was to screw her eyes shut and press her face into my shoulder. Her body was rigid with dread.

"Plus," I protested, "there aren't any words. You need to look to see the story." Flora said something I couldn't hear, and with a desperate hand gestured for me to read the thing, please, for pity's sake, and get it over with.

So I opened the front cover and told her about the hyperrealistic, dreamlike illustrations, which show a little girl coming home in the late afternoon. As she goes upstairs to her room, we see that she casts a shadow with a mind of its own. The shadow skulks and leers behind the child until, with a shock, she notices. Dropping her pencil and paper, she takes refuge behind a chair. The hunched shadow gets bigger, looming over the girl with horrid crooked fingers and flaming orange jack-o'-lantern eyes. Just as the shadow is about to get her, we see the child collect herself. Arms crossed, she stares it down. Instantly the shadow shrinks and cringes. The child points an indignant finger, and though there's no text to say so, it is clear that she's shouting something along the lines of "You stop scaring me, right now!" When the girl turns on a light, the shadow disappears. "Ta-da!" she seems to say, raising her arms in triumph. All terrors banished, she shows off her pencil drawings to her dolls. It's only at the very end, when the girl is fast asleep with her dolls held close, that we see . . . under her bed . . .

"Don't say it!" cried Flora, still hiding. "It's too scary!"

There was a pause as she mastered herself. Then:

"Please will you read it again?"

*   *   *

IN 2011 WRITER Adam Mansbach gave profane voice to frustrated parents everywhere with his not-for-the-kiddies picture book, *Go the F**k to Sleep.* Illustrated by Richard Cortes and written in the style of a gentle bedtime story, the book gave a generation of mothers and fathers gleeful permission to admit their exasperation with the nightly routine.

> *The wind whispers soft through the grass, hon. / The field mice, they make not a peep. / It's been thirty-eight minutes already. / Jesus Christ, what the f**k? Go to sleep.*

The book got an ecstatic reception. People found it hilarious and naughty. What looked like an attack on the final redoubt of cultural innocence, the tender bedtime ritual, was of course also a sly critique of ineffectual modern parenting. The narrator is doing what he's "supposed" to do, reading the kid a story, but the kid isn't buying it. He can tell that the narrator's heart isn't in it.

There is no question that some children are turbulent and hard to cajole. But persisting with *"Goodnight Moon* time," night after night, has cumulative benefits even beyond ultimately making it possible for parents to have a little time alone, like the exhausted, thwarted couple in *Go the F**k to Sleep*. Setting aside a substantial wedge of time before bed to read and talk helps to give shape and order to chaotic days. If children have been using screens it's especially important to create a calm hiatus between wakefulness and slumber. Their eyes and brains need time to power down. "Repetition and structure help children feel safe," advises psychiatrist Marie Hartwell-Walker; "bedtime declares that the day is over." In establishing a loving, predictable routine, she adds, "You are building your children's confidence in their world."

Routine is a gift for parents, especially first-timers who face a staggering learning curve. Proceeding from one preordained step to the next at bedtime according to a practical template—feeding, bathing, diapering, reading aloud—allows new parents to adapt to the shocking change in their circumstances. One great advantage of a newborn is that he has no idea of the uproar he is causing. While snoozing and feeding and snoozing and feeding, he gives his family members time to get adjusted to their new roles as his support staff (for a hilarious portrait of this plunge in status, see *His Royal Highness, King Baby*, by Sally Lloyd-Jones).

First-time adoptive parents may not get a grace period. If the child is already a few years old, they have to adapt without delay. Such was the case with Walter Olson and Steve Pippin when they brought their adopted son home to New York from an orphanage in Russia. Tim was three and spoke only Russian. It was a matter of urgency to get him settled, start him off with English, and forge those all-important emotional bonds. To achieve all three of these goals, Olson and Pippin put books and stories at the heart of an elaborate evening ritual.

"The first couple of weeks, bedtime was just terrible. He didn't want to go to bed at all, he fought it tooth and nail," Pippin told me when I visited the family in their current home in a small Maryland town. Sunshine filtered through chintz in the front windows, which admitted onto a narrow roadway that was once the gateway to the West.

"In the orphanage, it was like a barracks," Olson said, "and each kid's bed would be a few inches from the next bed over. There was one attendant who wanted to read her book, so was not interested in interactions with the kids."

When Tim arrived in the United States, his adoptive parents bombarded him with language. At first Olson and Pippin read aloud from Russian nursery books, having learned enough Russian to pronounce the words. Soon they switched to English. "Within two months, he was speaking half-English and half-Russian, and within six months he was really down to a countable number of Russian words, all the rest being English," Olson said.

"There were hold-on words," Pippin put in, "like *malaka* for milk, that stayed for months afterward. *Kasha* for cereal stayed for a long time, *baka* for dog stayed a long time, but there was a pattern where it would disappear from the vocabulary for a couple of days, and I don't know if that was just by accident or there was some kind of change going on, and then it would reappear in English and the Russian would be more or less forgotten."

The men devoted a solid hour to reading every night. Tim loved to hear the old-fashioned animal stories of Thornton Burgess, the early-twentieth-century newspaper columnist who wrote *Old Mother West Wind* and other books about characters with quaint names: Reddy Fox, Jerry Muskrat, and Jimmy Skunk. Tim's choice of books changed with time, but the nightly reading went on until he was about thirteen, Pippin said, "because he just *liked* it so much. It was a quiet time, it was communicating. You know, all of our bedtime rituals were basically ways of saying, 'Everything is fine, everything is the same as it was last night,' and the reading just put the cap on it."

As a toddler, Tim had been dropped into a new household in a foreign country with a strange language. He started learning English three years later than the American children in his classes at school. Yet soon he tested ahead of his peers in vocabulary. Coincidence? I don't think so. Hearing stories read for an hour each night had given Tim invaluable exposure to language forms and the pronunciation of English words. What worked for him, coming from Russia, can work for any child, from anywhere. Tim's experience validates the parental sacrifice that's required to create a regular bedtime ritual. It also demonstrates the power of an environment filled with language. Surround young children with lots of lovely words, it seems, and all manner of good things happen.

CHAPTER 5

THE RICH REWARDS
OF A VAST VOCABULARY

Babar is riding happily on his mother's back when a
wicked hunter, hidden behind some bushes, shoots at
them. The hunter has killed Babar's mother! The monkey hides, the birds fly away, Babar cries.

—Jean de Brunhoff, *The Story of Babar*

The most traumatic scene in a classic picture book had its
origins in the sweet tranquility of bedtime. In 1930, in a
house outside Paris, a young mother named Cécile de
Brunhoff made up a story while she was putting her two little sons
to bed. The boys were entranced by her impromptu tale of an elephant calf who is orphaned by "a wicked hunter." They urged their
father, who was an artist, to expand and illustrate their mother's
tale. Jean de Brunhoff obliged them. He put pencil to paper, exper-
imented with shapes and composition, and sketched out scenes.
Then, using ink, watercolor, and cursive lettering, he turned his
wife's story into a picture book that would become a global cultural phenomenon: *L'Histoire de Babar, le petit eléphant.*

Published a year later, the book we know in English as *The Story
of Babar, the Little Elephant* was followed by six sequels by Jean de
Brunhoff, and a whopping forty-five or so more by his son Laurent, who was one of the little listeners on the night Cécile made
up the original story.

In recent years, a certain amount of low-level controversy has attached to the Babar series. Some critics object to the first book's colonial aesthetic, with its implication that a wild elephant would prefer to be civilized, to wear a hat and suit, and to stand on his hind legs like a Frenchman. In one of the sequels, *The Travels of Babar*, de Brunhoff's depiction of cannibals strikes the contemporary eye as retrograde to the point of appalling. Then there is the famous shooting of Babar's mother, a scene so distressing that some children hide until the page is turned. The first Babar book is a bit weird, in truth, but there is no disputing its popularity. Translated into seventeen languages, the little elephant's adventures have sold many millions of copies.

Whatever you may think of the storytelling, there is also no denying the suggestive brilliance of de Brunhoff's illustrations. His pictures in *The Story of Babar* are crammed with detail and specificity, with action and objects and creatures. Every page brims with possibilities for conversation, for quizzes and questions, and for the chatty, dialogic give-and-take that can add so much to a child's bank of knowledge.

Consider the first page: "In the great forest a little elephant is born." There is Babar's mother, rocking her son to sleep in a hammock with the prehensile tip of her trunk. Mother and son are surrounded by green and yellow grasses dotted with red flowers. Tropical trees stand vigil, and two birds and a scarlet butterfly flit nearby. A single undulating line above the horizon marks a distant range of mountains. It is not a complicated picture, and yet it contains an amazing number of elements: a mother, a baby, a hammock, a trunk, a tusk, a butterfly, flowers, palm trees, birds, mountains, and the colors green, red, gray, and yellow.

For the baby or toddler on a parent's lap, any one of these objects might be new and captivating. ("Ah, so that is a "tusk," and, oh, those are "palm trees"!) *Babar* is full of depictions of commonplace things, such as cars and dogs and trees and birds. Yet it also abounds with illustrations of memorable oddities. Looking through the pictures, children will make the acquaintance, in a manner of speaking, of an opera house, a chandelier, a department-store floorwalker (wearing pince-nez!), shoes with spats, a pelican, a rhinoceros—even our old friend the andiron, peeking out from behind Babar's dapper form as he spins an after-dinner tale of his life "in the great forest" before he came to live in town with the wealthy and benevolent Old Lady.

Midway through the book, two pages show Babar motoring through the countryside in his shiny red convertible. The picture is a riot of dialogic possibility: in the distance, a tugboat puffs along with barge in tow, an angler reels in a fish on his line, a train chuffs over the arches of a distant bridge. There are cows, furrowed fields, blossoming trees, dragonflies and birds and insects and barnyard fowl; there's a hot air balloon and a church spire and a riverside restaurant and an airplane (with a propeller) overhead; there's a girl with her hair in a long single braid down her back, a goat with a bell hanging from a green collar around its neck, and a mile marker that resembles a small white tombstone beside the road. All this, in just two pages.

Elsewhere, more tantalizing tableaux: the Old Lady serves Babar soup from a tureen at her circular dining table. The two friends do their morning calisthenics (vocab word!), which Merle S. Haas, who turned de Brunhoff's French into English, charmingly translates as "setting-up exercises." Babar and his cousins Arthur and Celeste eat pretty pink cakes at a pastry shop, and the King of the Elephants tastes a bad mushroom, which causes him to turn green and die, an event that opens Babar's path to the throne. There is a merry dance after Babar marries Celeste, the couple's coronation as the new king and queen, and a final scene of the newlyweds beneath a serene and starry sky.

Everything I have just described appears in the pictures. Add the text, and the listening child will hear all sorts of other interesting and unusual words: fond, satisfied, elegant, learned (as an

adjective), becoming (ditto), progress, marabou bird, scold, promises, calamity, funeral, quavering, proposal, splendid, dromedary,
au revoir, honeymoon, and, "a gorgeous yellow balloon."

It takes just under seven minutes to read the forty-six pages of
The Story of Babar out loud, if you don't linger for quizzes. In that
time, a child will have vicarious emotional experiences. He will
see tenderness and catastrophe, fear and comfort, pride and anger,
death, marriage, sorrow, and joy. Such a profusion of image and
word and concept, and if you've set aside an hour for reading you
still have fifty-three minutes left. Think of the language riches a
child will acquire if this happens every day, starting when he is
tiny. His mind will become a hoard of glimmering, glinting, gem-
studded things.

*   *   *

"WORDS ARE AS wild as rocky peaks. They're as smooth as a mill-
pond and as sunny as a day in a meadow. Words are beautiful
things," said the writer Brian Jacques. Words *are* beautiful things.
They hold meaning, they reveal meaning, and they give us the
power to express meaning. Words are also keys that unlock the
world. Every time we read a book to a child, we are holding out a
new box of interesting and useful keys for them to collect: a tumble of shapes and colors in which they may discover vintage keys,
copper-colored pin-tumblers, tubular keys, double-sided keys,
grand brass lever-lock keys. The variety of the keys they find, by
its very existence, hints at the wideness of possibility in the world.

In medieval times, the lady of a castle, the chatelaine, could be
identified by the fact that she carried keys. With the keys and other
useful tools attached to a device strung with chains (also called a
chatelaine) at her waist, she could enter any chamber, any store-
room, any locked closet. Having the keys made her mistress of her
estate. The same is true for children and the words they learn: the
more they have, the more vaults they can open. Not only that, but

the more words they know, the more easily they'll pick up new ones from context, syntax, and repetition.

There is music and antiquity in our words. Ordinary language that you and I use with our children has come to us from the deep past, handed across generations through speech and print. Words are the raw materials of the "language arts," that stale phrase from elementary school that no more than hints at the emotional dynamism and potential for beauty we can unlock through near-infinite combinations of words. Language *is* an art form, if not always expressed in ways that exhilarate. It is also democratic and universal: anyone can dabble, and there are no expensive paints or canvases to buy.

"If your attitude to language has been generated by a parent who enjoys it with you," Philip Pullman has said, "who sits with you on their lap and reads with you and asks you questions and answers your questions, then you will grow up with a basic sense that language is fun. I can't begin to express how important that is; the most important thing of all."

Language allows children to occupy the world, their castle, as owners. It means they can understand and describe things with texture and precision. It means that if a girl sees a dog or a squirrel, say, moving with great speed, she can describe what's happening: is the creature darting or sprinting, racing or feinting, ambling or scampering? When something frightening happens, she can fine-tune her explanation: it was chilling, alarming, macabre, ghastly, daunting, or perhaps just unpleasant. Gradations of meaning matter, because they bring us closer to truth.

Even if nouns, verbs, adverbs, adjectives, and the rest had no practical application, it still would be good for children to cultivate an ample and varied supply of them. That they *do* have functional value makes the spreading of their goodness that much more important.

* * *

FACILITY WITH LANGUAGE improves a person's ability to succeed in the world, and it starts early. Young children whose heads are well stocked with words tend to enter school ahead of their peers. They start with an advantage that in most cases increases as they rise through the grade levels, because, in a ruthless natural calculus, vocabulary growth comes with its own built-in accelerant. As neurobiologist Maryanne Wolf explains in her book *Proust and the Squid*, "For word-rich children, old words become automatic, and new words come flying in, both from the child's sheer exposure to them and from his or her figuring out how to derive the meanings and functions of new words from new contexts."

Word-rich schoolchildren gather more words with each passing year, pulling ahead all the time as their word-poor comrades fall behind in comparison. Academics call this phenomenon of accumulated advantage the Matthew effect, after a line from the Gospel of Matthew: "For to everyone who has, more will be given and he will grow rich; but from the one who has not, even what he has will be taken away." In common parlance we might say, "The rich get richer, the poor get poorer," but in this case, the wealth we're talking about is language, which is free to all.

The University of Kansas researchers who uncovered the thirty-million-word gap two decades ago (the original report was entitled *Meaningful Differences in the Everyday Experience of Young American Children*) decided to revisit the original subject families for a 2003 report. What they found alarmed them, as was evident from their title, "The Early Catastrophe: The 30 Million Word Gap by Age 3." In describing their findings, Betty Hart and Todd Risley took pains to point out that all forty-two of the families they observed, whether tending toward word-richness or word-poverty, "nurtured their children and played with them. They all disciplined their children and taught them good manners and how to dress and toilet themselves. They provided their children with much the same toys, and talked to them about much the same things. Though different in personality and skill levels, the children all learned to talk and be socially appropriate members of the family with all the basic skills needed for preschool entry."

There was, however, no avoiding certain glaring differences in the amount and type of language to which the children of these families were exposed. In word-rich households, parents talked more and encouraged conversation through affirmations ("That's an interesting toy"), whereas in word-poor homes, the parents talked less and were more inclined to use prohibitions ("Don't touch that"). The children who heard more words learned more words. The children who heard fewer words learned fewer words.

Boys and girls who had entered the original study as seven-month-old babies had left it at the age of three with divergent abilities. The children from word-rich homes could muster about 1,100 words. The children growing up in word-poor households had access to around 500 words. These differences correlated with even wider differences a few years later. Having returned to the families in their first study, Hart and Risley were "awestruck," they wrote, to find that language competency at three accurately predicted measures of language skill *six and seven years later*, when children were nine and ten.

* * *

SO HOW DO we do it? How can we surround children with language in such a way that they will encounter a rich variety of words? How can we help them remember the words they hear? And how best can we encourage them to experiment with the words they learn, so that they not only acquire a wide vocabulary but can also use it?

We can start by reading out loud to them. Well, *of course*, you may be thinking. Yet the several ways that reading aloud helps a person develop a sophisticated vocabulary are not necessarily self-evident. It is worth looking at them in turn, and in detail.

To begin with, books contain words, and therefore when we read books out loud, we are delivering those words to the listener. So far, so good. But when the listener in question is a child and the books contain more pictures than words, how helpful is reading them, really? How many distinct vocabulary words can a child possibly hear?

The answer may surprise you. We know from the example of *The Story of Babar* that even a single picture book can contain multitudes of surprising and interesting words. It's true that Babar's story comes to us from the 1930s, a time when picture-book texts tended to be more discursive than they are today. Assuming that parents and children will draw from an array of books, old and new, what kind of language fuel are we talking about?

That's what researchers at Indiana University, Bloomington, wanted to find out for a study completed in 2015. The researchers began by selecting a hundred well-known picture books. These included beloved oldies—*Babar* and Munro Leaf's *The Story of Ferdinand* and Virginia Lee Burton's *Mike Mulligan and His Steam Shovel*—as well as popular newcomers such as *The Day the Crayons Quit*, by Drew Daywalt, and Jon Klassen's *This Is Not My Hat*. The research team measured the lexical diversity of each book—that is, the total number of unique words. Then they did the same with recordings of parents chatting with their children. The children in the recorded dialogues ranged in age from newborn to five years old, matching the intended age range of the picture books. This created clear points of comparison between the words that children would hear in everyday speech and those they would hear from books.

The result? The Bloomington team concluded that "shared book reading creates a learning environment in which infants and children are exposed to words that they *would never have encountered via speech alone.*" (Italics mine.) "Unlike conversations, books are not limited by here-and-now constraints," the report authors wrote, "each book may be different from others in topic or content, opening new domains for discovery and bringing new words into play." At the rate of one book a day, the Bloomington team found that "a child would hear more than 219,000 words of text in a year. At the rate of two books a day, the child would hear more than 438,000 words of text in a year."

For the average child, a diet of two picture books a day would supply about 6 percent of his "linguistic input," as it's called. That may not sound like much. But imagine expanding the daily reading from two books to, say, six or seven. Then imagine what this might represent in the life of a child who, like the word-deprived children in the Hart-Risley study, might otherwise be millions of words behind his peers by the time he turns three.

Every dollop of language can help. A word-poor household is, ipso facto, a place where not a lot of speech is directed toward children. It is often (but by no means always) low in socioeconomic status. According to a 2013 Stanford University study of low-income Latino families, most of the preschoolers under observation heard between 6,000 and 7,000 words of child-directed speech over the course of ten hours. That's not a lot. Yet there were striking variations between the twenty-nine families in the study. One lucky child had heard more than 12,000 words. A less fortunate toddler had heard a mere 670 words. Think of the boost a picture book or two would add to *that* child's repertoire.

What matters with vocabulary acquisition is not just the number of words but also their variety, and the ideas, objects, and concepts they represent. As New York University's Catherine Tamis-LeMonda pointed out, "When you look at the content of language that children hear in different settings, book sharing is really the *only* setting in which you could talk about things that are different from your everyday routines."

Picture books, she told me, provide "the opportunity to talk about many different words: the moon, the sun, the planets. You can do this in book sharing, whereas it would be kind of weird in everyday life. In everyday life, you may talk about books and balls and blocks and so forth, but books open the world and its infinite possibilities for words."

*   *   *

INFINITE POSSIBILITIES FOR words: it is a delicious idea! If only infinity were what small children wanted. Unfortunately, as many a fatigued grown-up will testify, often what's wanted is not a new story, with new words, but the same story over and over and *over*. It can drive you to distraction.

"She will ask me to read *Pip and Posy* again and again," said Magda Jenson, whose daughter was two when she made her complaint. "Eventually I say, enough! I've read *Pip and Posy* six times today! It's time to read something else!"

Her frustration is forgivable. It can be tedious to read the same book night after night. It may even seem unhealthy for a child to fixate on one story when so many others remain unexplored. If nothing else, our irritation marks us out as grown-ups. We love novelty; children love reassurance. We like smoky nuance; they like clear endings. In any case, Axel Scheffler's rabbit and mouse characters aren't supposed to interest people over the age of five. In bright pictures, the friends play and argue and experience toddler drama, and though adults may appreciate the books as pleasant bedtime reads, we cannot expect to be drawn into their thrall the way a two-year-old will be. So when the child holds up a *Pip and Posy* book yet again, we groan: "Enough!"

The trouble is, it may *not* be enough. When a child asks for the same story "again, again," he is telling us something important, though we may never find out what that important thing is. The book may be helping him perform quiet interior work having to do with fear or sadness that he can't articulate. The book may be an old friend whose familiarity feels comforting at bedtime. When

Flora was at the *Pip and Posy* stage, she wanted us to read Clement Hurd's *The Merry Chase* every night, for weeks. She loved the color-saturated illustrations of a dog dashing after a cat through a neighborhood—right through houses and shops—causing mayhem. She also had an inexhaustible appetite for Stephen Mitchell's adaptation of *The Tinderbox*, the Hans Christian Andersen story. Flora would pore over Bagram Ibatoulline's intricate pictures of the gnarled and terrible witch, and she wasn't troubled in the least when the three huge dogs ("the one with eyes like clocks, the one with eyes like dinner plates, and the one with eyes like wagon wheels") put a violent end to the king and queen.

Why did she love these books so? I don't know. I don't have any explanation, either, for Molly's devotion, years earlier, to *Around the World with Ant and Bee*, by Angela Banner, or to David McKee's cautionary tale *Not Now, Bernard*. All I know is that we went through these books again and again and again. I noticed, as many parents do, that my children had little habits with their favorite books, moments when the page had to be touched in a way that could appear oddly formal. If we had once paused to examine a tiny drawing of Lowly Worm in his little Tyrolean hat in a busy Richard Scarry illustration, for instance, we had to pause every time. If a page had a rip or a smudge, a small finger might have to touch that spot, too, every time. One little girl, Ella, who's all grown up now, was so enamored of the dogs splashing in a swimming pool in a P. D. Eastman illustration that she would try to join them. Her mother laughed at the memory: "When we got to that part, she had to take her sock off and put her foot physically into the gutter of the book, as if *she* were going into the pool. I remember being on the subway with the stroller and she'd be strapped in, and we'd get to that page and I could see her, in the stroller, physically trying to get her foot up so that she could get it into the book!"

When my daughters Violet and Phoebe were four and three,

respectively, we had to read *The Story of Babar* most nights. Violet at the time had a curious cosmology: in picture-book illustrations, all older men were kings, all younger men were princes, and females of all ages, unless obviously witches, were princesses. In those days my children also competed to claim characters and objects in the picture books we read ("I'm the monkey," someone would say, tapping the page, or "That's my cake"). One night the girls and I arrived at the scene in which Babar meets his wealthy benefactress. Violet thrust out a finger and pointed to the Old Lady.

"I'm her," she said. "She's a princess."

A few pages later, we got to the scene of Babar motoring through the countryside, about to pass the girl with the braid down her back standing beside the goat with the bell around its neck.

It was Phoebe's turn to poke the page.

"That's me," she said, "with my goat."

"No, that's me," Violet said.

"No, that's you," her little sister agreed.

"I know," Violet said. "She's a princess."

*   *   *

WE MAY NOT ever know why some books come to exert such spellbinding power that children want to hear them again and again. Perhaps it will forever be a mystery, like love. There does however seem to be one solid, prosaic explanation: children enjoy repeating books because the experience imbues them with feelings of competence and mastery; because, with each reading, they understand a bit more of what they're seeing and hearing.

Researchers at the University of Sussex in Great Britain tested this idea by exploring the effect of rereading the same storybooks on a cohort of three-year-olds. I should point out that the books they chose did not reflect the passions of any particular child but were created for the study. The clinicians wanted to observe the children's fast and slow mapping of new words, that two-speed process of language acquisition we discussed earlier. So they smuggled a couple of made-up vocabulary words into the narratives of otherwise unremarkable picture books, so that they would be able to separate the ersatz words from words the children already knew. For three specially created picture books, *The Very Naughty Puppy, Nosy Rosie at the Restaurant,* and *Rosie's Bad Baking Day,* the researchers invented an unorthodox hand mixer, the *sprock,* and a peculiar rolling pin, the *tannin.*

To the young subjects, these words would be no more bizarre than the aardvark or the andiron. The English language is full of eccentricity. Why not a sprock or a tannin? The children in the study took the words in stride as they would any others that were unfamiliar to them. Context gave them clues to the meaning of the words and their grammatical function, and repetition made the words memorable.

As Vanderbilt's David Dickinson and his colleagues pointed out, "Children learn vocabulary through grammar and grammar through vocabulary." Once they know a word, if they hear it in different syntactic settings, their understanding will expand. To take an example from *Babar,* the adjective *becoming,* to describe a shade of green, turns into a verb when used to describe the changing of one circumstance into another ("it's becoming chilly"). The more words and the greater the diversity of texts children hear, the more easily they can untangle these intricacies.

The Sussex University study found that reading picture books "again, again" is remarkably helpful in this regard. "We found a dramatic increase in children's ability to both recall and retain novel name-object associations encountered during shared storybook reading when they heard the same stories multiple times in succession," Jessica Horst, Kelly Parsons, and Natasha Bryan wrote, publishing their conclusions in 2011.

What's more, the children who heard (and saw) repeated iterations of the words in the same stories retained the new words to a much greater degree than those who encountered the words (and objects) spread across different stories. The results, the researchers wrote, "provide good news for parents: it is not necessarily the number of different books that matter, but rather following requests to 'read it again!'"

That will be some small consolation, I hope, if a child presents you, for the hundredth time, with a well-worn copy of *Pip and Posy*.

* * *

THERE IS A third important way that picture books foster an atmosphere rich in words. When an adult and a child sit together and leaf through pages filled with writing and artwork, it is a natural and easy time to talk. If the child is very young, the "conversation" may be one-sided and fairly primitive but that doesn't mean it isn't valuable. It is. Every scrap of informal chatting that we do over picture books is fuel for the engine of language acquisition.

Academics who study the subject often cite a little fable known as "three mothers and an eggplant." It is not a fairy tale—the eggplant doesn't talk or grant wishes—but what it teaches, the moral of the story, is helpful in understanding how small interactions can produce big increases in a child's vocabulary. The fable takes place in a supermarket:

The first mother wheels her shopping cart down the produce aisle, where her kindergartner spots an eggplant and asks what it is. The mother shushes her child, ignoring the question.

A second mother, faced with the same question, responds curtly, "Oh, that's an eggplant, but we don't eat it."

The third mother coos, "Oh, that's an eggplant. It's one of the few purple vegetables." She picks it up, hands it to her son, and encourages him to put it on the scale.

"Oh, look, it's about two pounds!" she says. "And it's $1.99 a pound, so

that would cost just about $4. That's a bit pricey, but you like veal parmesan, and eggplant parmesan is delicious too. You'll love it. Let's buy one, take it home, cut it open. We'll make a dish together."

So: three mothers, three distinct responses to a child's simple question. The women's replies help to explain why some children come out of toddlerhood knowing lots of words and concepts, and some do not. The first mother doesn't engage the child at all. The second mother acknowledges the question, but shuts down further conversation. The third mother uses the query about an eggplant as a starting point for a disquisition on all things aubergine: the vegetable's color, its weight, its price per pound, its flavor compared with that of a dish the child has tasted, and its relevance in the family's life and diet.

What we see, with the third mother, is a kind of a three-dimensional, book-free version of a practice known as interactive reading or dialogic reading. Asking and answering questions, seeking and finding things in pictures, riffing on language, fooling around with alliteration or rhymes; all these are dialogic techniques. It's like a form of play. And, as it happens, research tells us that children given the chance to hear and use vocabulary in a playful setting remember it far better than those who get straightforward instruction.

As Roberta Michnick Golinkoff, professor of education, psychology, and linguistics at the University of Delaware, told me, "The child learns best when they're active, not passive. But you don't want to turn reading into didactic teaching time. You want to follow the pointing finger, the little pointing finger, so that what's on the page comes off the page and links up to the kid's life."

Even very young children will give clues about what interests them. They may whack the book, and bend it ("Hmm, what are the properties of this physical object?"). They may want to turn the page, or indicate the faces of animals or people, or, with that little pointing finger, trace the outline of a shape or a letter. The more they "read," the clearer the hints they will give. That's where the grown-up comes in, following the clues and improvising accordingly.

Caroline Rowland, professor of psychological sciences at the University of Liverpool in Great Britain, explained how dialogic reading should evolve. "If you're reading with a one-year-old, you probably want to do factual stuff—you know, 'Ooh, look at the doggie, can you see the doggie?' and waiting for the child to point at something and then talking about what they're pointing at and describing it," she said. "But as your children get older, three to four, then you might want to change your interaction so that you're doing more decontextualized talk." This means unhitching the element of the story in question and connecting it to the child's experience of the world.

"If there's a doggie in the book, you can talk about the doggie you saw in the park, and that there are lots of different types of doggies," Rowland went on. "The great thing is that you're *both* focused on one thing, you *know* the child is interested because they're engaged, and you can adapt your talk to whatever is going to be most effective given the age of your child, and how much language your child already knows. That's one of the reasons it's so effective. You can be very sensitive to your children's developmental levels, and can keep boosting them a little bit further from where they are."

The idea is to encourage conversation and interaction in a way that everyone enjoys and that offers bit of fun and challenge, like our old game of quizzes.

In themselves, picture books are objects worth talking about. There's the front cover, to start with: What does it show and what feelings does it elicit? The endpapers may be designed and decorated so as to establish a mood or to plant an idea before the story begins. That's worth talking about, too. Then there's the story itself, and the illustrations, and the limitless possibility of things to wonder and say about them. A parent might linger over a picture and ask the child to find everything that is red, or square, or liquid; to name the parts of the body, or identify different pieces of fruit. A mother might teach her child the names of animals in both English and Japanese (assuming she studied it for longer than I did) or in Spanish or Tagalog or Korean. The two could practice counting objects up to ten together, or counting back down from ten to zero. A father could ask a child to find objects that are "on top of" or "inside" or "underneath" other things, to develop an awareness of spatial concepts. To stretch a child's vocabulary, adults might use baroque description, as I liked to do, to add a layer of decoding to the process of locating and identifying objects in the illustrations.

Many picture books come with the principles of dialogic reading already built into them. Often questions will be tucked in amid the illustrations: "Which mountain goat is happy?" or "Whose baby is getting a bath?" All a grown-up need do is open the book, say the words on the page, and presto: he or she is expanding a child's world and his ability to understand and describe it.

As productive as talking may be, sometimes a moment of quiet is in order. One father told me that he likes to stop reading now and then to give his young son a chance to think about the story, and perhaps to comment. Open-ended questions are good, too. Picture books such as *The Tale of Mrs. Tiggy-Winkle*, by Beatrix Potter, Virginia Lee Burton's retelling of *The Emperor's New Clothes*, or Mac Barnett's *Sam and Dave Dig a Hole* give ample opportunity for the child to come up with his own thoughts and opinions. Parents can encourage this along with an occasional, gentle "I wonder why?" or "What do you think about that?"

One woman told me that when her two children were small, she used story time to impart deliberate lessons. In particular she wanted to help them develop a skill known as auditory discrimination. It is the ability to distinguish different sounds—the subtle contrast between the sounds of "t" and "d," for instance—that children need when it comes time for them to learn to read on their own. The woman told me that she would sit in a rocking chair with the toddlers on her lap. "I'd say, 'Can you find something in this picture that begins with the same sound as the word "dinosaur"?'" Sometimes she had them clap when they heard a specific sound, which they thought was hilarious. Over time, using these techniques, she taught them initial consonant sounds, went on to initial blends (that is, combined consonant sounds such as "bl" or "sk" at the start of words), and even led them on to the confusing terrain of long and short vowels. Without the children realizing it, their mother was equipping them for preschool. From their point of view, they were all just having a happy time together before she tucked them into bed.

*   *   *

WHAT HAPPENS WHEN children are old enough, and adept enough, to read to themselves? Well, then they can read to themselves, and that's terrific. But it doesn't mean that reading aloud somehow fades in relevance. On the contrary: it retains its many powers—to enrich, enlighten, transport, and transform. To me, the serious joy of the thing begins when reader and listener can meet each other in substantial, demanding stories that will repay the effort required to read them a thousand times over in richness of language, of character, and of lasting imaginative effect.

I do want to say a word here about effort. It takes time but is otherwise not very taxing for most adults to read a handful of picture books, though, heaven knows, there are nights when even one short book feels like an imposition ("Go the f**k to sleep!"). Reading every day, or close to it, takes discipline when children are little. It takes a real act of will as they get older and other claims begin to encroach on the time they have at home. Schoolwork, sports, friends, part-time jobs, and the hydra-headed temptations of technology will try to crowd out regular reading. Don't let it. This is a battle worth winning.

Children and parents can have full lives in cyberspace and in unplugged reality *and* find time to meet in literature. We don't have to give up our devices. If anything, reading together allows us, and our children, to live in easier harmony with our machines. It gives us a time every day when we can reconnect in a low-pressure way and enjoy a little bouquet of neurochemicals, even as the books we read fill our children's minds with ever more sophisticated language. Making the time to read together is almost an obstinate act of love. The mutual effort—the sacrifice of time—becomes part of the reward.

Stick with it, and the compensations are extraordinary. The baby girl who lifted the flaps of Rod Campbell's *Dear Zoo* becomes the toddler charmed by Ludwig Bemelmans's *Madeline*, who turns into the sixth-grader listening openmouthed to Mark Helprin's *A Kingdom Far and Clear*, who grows up to be the young woman swept away by Leo Tolstoy and the beautiful, ill-fated heroine of *Anna Karenina*.

Each book makes straight the path for the next, opening out into sunlit literary meadows where, over time, young people will encounter beautiful writing and characters and scenes that may have been loved, known, and remembered by generations long since passed. For the child or teenager (or anyone else, for that matter), getting these tickets to Arcadia is a matter of simplicity. All they have to do is listen.

Most of us understand far more words than the ones we use in daily speech. We know this is true from the example of babies, who can show that they understand simple language long before they have the power to engage in it. We know it from children who are late to speak. For that matter, even dogs have shown under MRI scanning that they can comprehend certain words, regardless of the speaker's tone of voice.

A child's receptive vocabulary, the words he can understand, is thought to be anywhere from one to three years ahead of his expressive vocabulary, the words he can use. This means that by ear he can grasp and appreciate narratives that would otherwise be outside his scope of competence. As Jim Trelease, author of *The Read-Aloud Handbook*, points out, a child's reading level doesn't typically catch up to his listening level until about the eighth grade. An adult reading aloud does far more than impart a story, therefore: he or she also shows by tone of voice, phrasing, and pronunciation how complicated sentences can be tackled, subdued, and enjoyed. And while all that is happening, the child is soaking up fresh ideas and unfamiliar words.

"Students don't learn new words by studying vocabulary lists. They do so by guessing new meanings within the overall gist of what they are hearing or reading," observed E. D. Hirsch, a former professor at the University of Virginia who is perhaps best known for his 1987 bestseller *Cultural Literacy*. "And understanding the gist requires background knowledge. If a child reads that 'annual floods left the Nile delta rich and fertile for farming,' he is less likely to intuit the meaning of the unfamiliar words 'annual' and 'fertile' if he is unfamiliar with Egypt, agriculture, river deltas and other such bits of background knowledge. . . . Vocabulary-building is a slow process that requires students to have enough familiarity with the context to understand unfamiliar words. Substance, not skill, develops vocabulary and reading ability—there are no shortcuts."

As a child is collecting words, he's also picking up usage norms and the approximate rules of the grammatical road. "There's a hidden form of vocabulary for kids when they're reading, that is syntactic complexity," educator Doug Lemov recently told an interviewer. "When you ask kids about a difficult passage, often times they got one of the ideas within a sentence, but the sentence was incredibly complex and multifaceted and so they didn't understand how all the ideas in the sentence connected. There's too much syntactic complexity for them."

Lemov went on to describe his experience of reading a well-known novel by Scott O'Dell to his young daughter. "There's nothing that she can read to herself that is like the complexity and depth of the narrative of *The Island of the Blue Dolphins*, which is just truly a great novel," he said.

> So I think that I'm selling her on the act of reading books, great books, by reading something that's beyond anything that she's ever imagined a book could do, and I think that for the rest of her life she will be changed by the experience of that book.
>
> The key to me is that I was reading her a variety of ornate sentences that are probably more advanced than most adults use in their everyday work lives, hundreds or thousands of them in a row, as a second-grader. And by expressing them, I helped her to understand what they sounded like, which is why there were comparatively few moments of lack of understanding.

This phenomenon opens the way to one of the most exciting and underappreciated satisfactions of reading aloud: that you can, as a parent, share intricate, powerful stories with your children, and *they will understand*. You can see it—the clouded, uncertain look of the child who is tuning out and a little lost, for sure, but also the strange energetic gleam in the aspect of a child who has just stepped out into one of those sunlit meadows. It is stupendous. You can't see what he's seeing, but you can tell that he's enchanted.

A capable reader of eleven or twelve might have a real struggle making his way through the subordinate clauses of a nineteenth-century novel, if he tried it on his own. Yet let that same child relax in an armchair while someone else takes on the text, and he's free to experience what becomes a seamless and thrilling whole. Vocabulary, syntax, plot, characterization: all these dry ingredients combine to form a rich and immersive experience. And if there is a moment's confusion, a word mistaken, or an idea that needs explanation, you can pause to supply what's needed. That's the beauty of contingency both when a child is small, during dialogic reading, and when he's older, and the two of you are taking on bigger books.

Not long ago, Flora asked me to read Bram Stoker's *Dracula* to her. Told through letters, diary entries, and newspaper clippings, the story is full of complicated and obscure language (not surprising in a novel first published in 1897). I was prepared for her to lose interest. She didn't. She was gripped. At the age of eleven, she took the archaisms in stride, and she did not seem to mind or even notice passages that struck me as convoluted or windy. With the young lawyer, Jonathan Harker, she met Count Dracula in his looming castle. She traveled with the monster, sealed in his repulsive box filled with earth, on a doomed cargo ship bound for England. She listened, rapt, as the count fed at the throat of Lucy; visited his mad disciple, Renfield; and attacked Mina, Jonathan's bride. As the story swept on toward its end, Flora rallied with the forces of good—Jonathan, Mina, Quincey Morris, Dr. Van Helsing, Lord Godalming, and Dr. Seward—as they traveled to Transylvania, formed three parties, and fanned out across the countryside, determined to stop Dracula from regaining his castle.

Flora drank in every word, and she would have listened for hours every night, rather than just the one I could give her. As we neared the end, though, she began to worry. She'd been around books long enough to suspect that Bram Stoker would kill off at least one of his heroes in the final confrontation. Who would live and who would die? She chewed her fingertips.

"I know it's going to be Quincey! Because he's so nice, but he's not as important."

She fell silent for a moment, and then gasped. "And Van Helsing! Because he's old! And he said he was willing to die for Mina!"

We grimaced at each other—the suspense was killing us—and I picked up the book again. Flora leaned forward in her chair, as if trying, like that little girl, Ella, to climb bodily into the story. Now Mina Harker and Dr. Van Helsing were making camp by themselves in the Carpathian wilds. In a horrible development, Count Dracula's three female companions materialize near them in the gloom. The horses are screaming with fright and tearing at their tethers, but the two travelers are safe within a protective charm. By the time dawn breaks, the spectral visitors have vanished, and Mina is fast asleep.

"'I fear yet to stir,'" Van Helsing says in his imperfect English. "'I have made my fire and have seen the horses; they are all dead. Today I have much to do here, and I keep waiting till the sun is up high; for there may be places where I must go, where that sunlight—'"

"What is he going to do?" Flora cried, jumping up. "Is he going to kill Dracula? Where's he going to go?"

"I don't know!"

"Please keep reading!"

It took me a moment to find my place again, at which point I resumed my feeble imitation of a Dutch accent.

"'—where I must go, where that sunlight, though snow and mist obscure it, will be to me a safety. I will strengthen me with breakfast, and then I will to my terrible work. Madam Mina still sleeps; and God be thanked! She is calm in her sleep . . .'"

Two paragraphs later, the other two parties of good guys—Jonathan Harker and Lord Godalming from one direction, and Dr. Seward and Quincey Morris from another—come within sight of the convoy of Romany bodyguards transporting the undead count in his sinister box.

"'With the dawn we saw the body of Szgany before us dashing away from the river with their leiter-wagon,'" I read, realizing a split second late that I had got the emphasis wrong. I had

pronounced *body* as if it were a dead thing, rather than a comple-
ment of men.

I glanced at Flora. "Did you get that bit?"

"I think so, but whose body was it?"

I'd been right. "'The body of Szgany' in this case is not one person, but a troop of men who are called the Szgany. You can have
a single body—let's say a dead body. You can also have a group or
assembly that also goes by the collective noun *body*."

"Oh."

"For instance, the United Nations is a deliberative body, and the
US Senate is a legislative body."

"Got it. Could you—"

The pedagogue was just warming to her work.

"So," I went on, "you could talk about a person having a body
of work. For example, we could say that in Bram Stoker's body of
work, the novel *Dracula* is—"

"Okay!" Flora was jumping in her seat again. "That's good to
know, but can you please keep reading!"

We both laughed, and I resumed: "'With the dawn we saw the
body of Szgany before us dashing away from the river with their
leiter-wagon. They surrounded—'"

"A what-wagon?"

"Leiter-wagon. Long, wooden, not very heavy," I guessed (cor-
rectly, as it happens). "'They surrounded it in a cluster, and hurried
along as though beset. The snow is falling lightly and there is a
strange excitement in the air . . .'"

From that point to the cataclysmic ending eight pages later,
neither of us interrupted. We gaped as Van Helsing thumped
stakes into the hearts of the count's female companions, freeing
their souls. We winced as Quincey Morris took a mortal Szgany
knife-thrust just as Flora predicted. We held still as, in a single
sentence, Dracula's assailants pierced his heart, sliced off his head,
and watched his body crumble into dust.

When I finished reading, there was silence. Flora looked shocked and regretful. The story was over.

We had been reading *Dracula* every night for a couple of weeks, now side by side, now settled a short distance from each other, always close together during the scariest bits. Flora had rattled in a carriage through the Romanian countryside, wondering as villagers flashed the sign of the evil eye. With Jonathan Harker, she had noticed the count's long fingernails, the hair on his palms, his ability to creep up walls like a lizard. She had stowed away on the Russian ship *Demeter* and, with its appalled captain, seen the crewmen disappear one by one. She'd visited poor Renfield behind the bars of his cell, feeling horror and pity as he ate flies and spiders and waited for his "master" to arrive in England. She had seen Lucy vivid, then languishing, then restored and ferocious in vampire form, and she had seen Lucy's friends administer a terrible cure with blade and garlic.

Flora had, in short, run the full course of Stoker's famous gothic horror story, and she had loved every page. It's not a book all eleven-year-olds (or all fifty-year-olds) would enjoy, but again, that's one of the satisfactions of reading aloud at home: to each his own.

Flora's expression was still dismayed. Then her face brightened. I should have known what was coming.

"Again!" she cried. "Let's read *Dracula* again!"

CHAPTER 6

THE POWER OF
PAYING ATTENTION—AND
FLYING FREE

———•═══•———

Alice started to her feet, for it flashed across her mind that she had never before seen a rabbit with either a waistcoat-pocket, or a watch to take out of it, and burning with curiosity, she ran across the field after it, and fortunately was just in time to see it pop down a large rabbit-hole under the hedge. In another moment down went Alice after it, never once considering how in the world she was to get out again.

—Lewis Carroll, *Alice's Adventures in Wonderland*

With Alice's dash down the hole under the hedge, Lewis Carroll captures what happens when a listener disappears into a story that someone is reading out loud. He doesn't know where he's going or what he'll find. His adventure may begin, like Alice's, with a momentary scramble. There's a book to be chosen, a seat to be claimed. As he settles in to listen, his surroundings are still ordinary, just as Alice, pursuing the white rabbit, travels along what at first appears to be a normal burrow that runs straight on, "like a tunnel." Then without warning, the rabbit hole dips down, and Alice finds herself falling into what seems to be a very deep well.

"Either the well was very deep, or she fell very slowly, for she had plenty of time as she went down to look about her, and to wonder what was going to happen next," Carroll writes, as if he were deliberately trying to reproduce the experience of a reader dropping into a fictional world, entering what novelist John Gardner calls a "vivid and continuous dream."

Everyone loves a good story. As the folklorist Sybil Marshall observed, "It seems that mankind is born with an avid appetite for details of other lives beside the one his own small span of corporeal existence grants to him; it is as though he seizes from his earliest years upon this way of enlarging the bounds of his own life."

Something special happens when this fictional transport takes place in the intimate setting of a read-aloud. The listener enters a cycle of thought, imagination, and practical behavior that can have surprising and even profound ramifications. Rather than taking prefabricated entertainment off a screen, he brings the smells, sounds, and sights of a story to life in his own mind.

In the view of the psychologist James Hillman, oral storytelling is not only good for the soul but also prepares children for life. Those who experience it in childhood, he argued, "are in better shape and have a better prognosis than those to whom story must be introduced. . . . Coming early with life, it is already a perspective to life."

In literature, we are freed from physical constraints and from the orthodoxies of our time and place. We meet characters we would never encounter in the real world. In a vicarious way, we experience life through them, and one result is an expansion of emotional understanding. As Britain's former children's laureate Chris Riddell said, "A good book is an empathy machine."

Complicated and mysterious things happen inside people when we give them time to listen. The trick is to make it happen.

*   *   *

ON A RECENT winter day, I got to watch a roomful of fourth-graders teleport to medieval Persia. The mechanism was a picture book by Diane Stanley. The medium was the voice of a school librarian, the celebrated writer Laura Amy Schlitz. The launch site, so to speak, was a Baltimore school library, a midcentury modern place of exposed brick and blond wood with a large window at one end that showed a bleak and overcast sky.

I had only just arrived and shaken hands with Schlitz when there was a sudden rush and bubble of voices from the hallway and in came a horde of nine- and ten-year-olds. Pouring past us, the children threw themselves on a terrace of upholstered benches, chattering and jostling. After a moment, without saying a word, Schlitz planted a yardstick on the floor with one hand and raised the other, rhapsode-style. The talk died away. These children were evidently in the habit of being interested in what their librarian had to say.

Schlitz pulled up a chair to face the class, made a few preliminary remarks, and held up a copy of *Fortune*. The front cover shows a man in a turban and robe gesturing toward an elegant woman who has knelt to embrace a tiger. Flowers and curlicues decorate the margins in the style of an illuminated manuscript.

"'Long ago, in the poorest corner of Persia, there lived a farmer and his son, whose name was Omar,'" Schlitz began. "'When Omar came of age, all his father could give him were his blessing and a small purseful of money. With that he would have to make his way in the world, but poor Omar had no idea what to do or where to go . . .'"

Apart from the single voice, the room was silent. The children's faces had slackened. One boy had pulled his knees up under his shirt so that only his feet peeked out. A girl sat cross-legged, leaning forward on her elbows. Another boy lay on his back, gazing at the ceiling. Everyone was listening.

"'You there, young man!'" Schlitz said, giving an imperious edge to the voice of a veiled stranger. The woman proposes to sell Omar a tiger on a leash: he can make the animal dance on its hind legs for money, and secure his future. Omar agrees.

"'Sure enough, everywhere he went with his tiger he was showered with money. And with the beast curled up beside him at night, he never feared robbers . . .'"

A teacher's aide reached across and raised a boy who had folded over. Otherwise, the listeners sat still. Their faces were blank. They had gone to Persia.

In the story, Omar gets very rich thanks to the tiger, and he decides that it's time for him to marry. Wealth has made him conceited, though, so Omar resolves to find a fancier bride than his childhood sweetheart, who's named Sunny.

> "You are a fine friend," he said, meaning to be kind. "You are not ugly, but you are not pretty, either. You are a farmer's daughter, you see, and I am now a great man. I should probably marry a princess."
>
> And with that he left his village and his little friend Sunny, whose sad, dark eyes reminded him strangely of the tiger's.

Soon Omar enters a magnificent city known for the sorrow of its princess. No one has been able to console Princess Shirin since her fiancé disappeared on the eve of their wedding.

"'Most people said he had drowned,'" Schlitz read in the voice of a gossipy serving woman who's explaining the situation to Omar, "'but some believed he had been enchanted by a witch.'"

I was impressed that Schlitz wasn't turning the story into a pedagogical exercise. She didn't stop to ask the children to predict plot turns or to analyze the characters. She didn't even turn the book around every few minutes, in that well-meaning but cumbersome way, to ensure that everyone got to see the pictures. She just let the words float out, one after another, trusting the language to work its own magic.

"Ooh!" gasped a curly-haired boy. His face brimmed with glee as he looked around to see if anyone else had realized something important about the tiger. No one had. He was the first to see that the animal is really the princess's true love, under a spell. Silly Omar, thinking he could woo the young woman, has instead brought her fiancé back to her. Soon—shazam!—the tiger disappears, and in its place stands a handsome young man.

Everyone thinks Omar has brought about the reunion on purpose. The grateful sultan heaps him with treasure. Only Omar knows the depths of his own foolishness. Shamefaced, he slinks back to his village to apologize to Sunny, and to ask her to marry him. "Well," says Sunny, "I don't know. You're not ugly, but you're not handsome, either. And you're certainly not a prince."

A ripple of laughter ran through the children. A moment later, Sunny relents and the couple weds. The class applauded. They were back in Baltimore. It had been a good trip.

*   *   *

THE EPISODE WAS an object lesson in the power of a good story to hold people fast. While Schlitz was reading, the boys and girls appeared to be in a state of suspended animation. They would not have realized it, but the enjoyment they got from listening was feeding a virtuous behavioral circle. If they sat still, kept quiet, and paid attention, they could enjoy the story, even as their enjoyment of the story got them used to sitting still, keeping quiet, and paying attention. An illustrated tale of a tiger, a foolish fellow, and a sorrowful princess had the meta-effect of encouraging these fourth-graders to extend the span of their attention.

That is no small achievement. In our distracted age, it's a challenge to keep anyone's focus. Technology is training us to dart and react like hummingbirds, scrolling, clicking, tweeting, liking. For people in the sustained-attention business, not least book publishers, these developments are unsettling. As Carolyn Reidy, the president of Simon & Schuster, worried aloud to an audience of booksellers not long ago, "You have whole generations being trained for shorter attention spans than books require."

The pleasures of technology produce a faster and more frenetic behavioral circle. Scrolling, clicking, liking, and tweeting deliver tiny doses of pleasing chemicals, which fuel our urge to scroll, click, like, and tweet again—a cycle that, Adam Alter points out, makes our devices almost irresistible. One of the effects is a decreasing ability to stay on task. It is said that since the turn of the millennium, the attention span of the average adult has dropped from twelve seconds to eight. If that's true, then Microsoft CEO Satya Nadella is right in saying that human attention is becoming "the true scarce commodity."

As grown-ups, we can decide for ourselves whether a longer attention span is something we want to take the trouble to maintain. With children, the matter is more serious. They *have* to be able to pay attention at school. And they need to understand what's being said to them, especially in the younger grades, when, as Jim Trelease has pointed out, most instruction comes through speaking.

When it comes to paying attention, children from read-aloud families go to school with a triple advantage. They're used to listening, so it's easy for them to do it. They've heard lots of language, so their comprehension will be comparatively strong. And they know from experience that paying attention brings rewards. These assets are not trivial. Studies have uncovered a strong correlation between the capacity of children to attend when they are small and their ability to do well in math and reading when they are older (it's a link reminiscent of the one between early language and later math, discussed earlier). In 2013, researchers at Oregon State University found that the "attention-span persistence," as it's called, of four-year-olds predicted their math and reading achievement at age *twenty-one*. Not only that, but age-four attention-span persistence also foretold whether children would finish college by the time they hit twenty-five. So there can be real consequences if children do not develop their capacity to listen at length.

*   *   *

AFTER LAURA AMY SCHLITZ had finished reading to the fourth-grade class, I had a chance to sit down with the children. I wanted to know what they made of the experience. How did it feel? What did they think?

Their answers gave a wonderful glimpse into the secret work of the imagination, and the gratification of engaging one's capacities.

"When Laura reads out loud to me, my brain shuts out everything else."

"I feel excited and interested."

"I feel like I am the main character when someone reads aloud to me."

"When someone's reading to me, I can do deeper thinking about things, like wondering if the tiger was the prince."

"I can be absorbed if I'm reading by myself, but I notice that it's easier when someone else reads, so my brain doesn't have to split its attention to the words and what they mean."

"I feel like I'm solving a mystery in the pauses. I forget that I'm right here. It feels like I'm in the sky, looking down from above, and can see the whole story."

This last feeling, the out-of-body sensation of drifting above events, or within them, or being swept away by them, is both common and extraordinary. The annals of personal remembrance and literary memoir are full of these transporting moments.

Walter Olson (father of Tim, the toddler adopted from Russia) was a little older than the children in Laura Amy Schlitz's library class when he first had the experience, and it never left him. In fifth grade, his teacher took out a full-length biography of the Indian warrior Crazy Horse and began reading it aloud to her class.

"This went on for weeks," Olson remembered, "and I was mesmerized. *Oh, my goodness, she is actually going to read us this entire book!* It was a luxury like I'd never known before, to absorb an entire book by having someone clearly and expressively read the whole thing to me."

Bouts of asthma kept the writer Alberto Manguel confined to his bed when he was a child. Propped up on pillows, he listened to his nurse read the "terrifying" tales of the Brothers Grimm. "Sometimes her voice put me to sleep," he remembered, "sometimes, on the contrary, it made me feverish with excitement, and I urged her on in order to find out, more quickly than the author had intended, what happened in the story. But most of the time I simply enjoyed the luxurious sensation of being carried away by the words, and felt, in a very physical sense, that I was actually traveling somewhere wonderfully remote, to a place that I hardly dared glimpse on the secret last page of the book."

As a second-grader, the novelist Kate DiCamillo "lived," she has said, for the times when her teacher read *The Island of the Blue Dolphins* to the class. Reading aloud "can change a child's life," DiCamillo said. "It can turn a child into a writer. It can certainly turn a child into a reader."

For the French author Daniel Pennac, the experience looks like liberation. In his book *The Rights of the Reader*, he writes, "Free. That was how our children experienced it. A gift. Time out. In spite of everything. The bedtime story relieved them of their daytime burdens. Freed from their moorings, they traveled with the wind, infinitely lighter. And the wind was our voice."

A gift, a luxury, a life-changing freedom: Why would we trade it for anything? Why would we stop? Yet in most households, even when parents have read to the children when they are young, that is what happens. Reading aloud begins to flicker out by the time kids turn five, according to Scholastic's biannual surveys. For the vast majority of nine-to-twelve-year-olds, story time becomes rare to the point of nonexistence. Yet in the 2014 survey, 40 percent of children between the ages of six and eleven said they wished their parents had not stopped reading to them.

They wanted to keep flying.

*   *   *

IN THE OLDEN days of the twentieth century, a correspondence school for "speedwriting" ran magazine ads that read:

> f u cn rd ths, u cn
> bcm a sec & gt
> a gd jb w hi pa.

My friends and I laughed over the ads. We could read the ad copy. We could become secretaries and get good jobs, with high pay!

The point was, of course, that we didn't need to be given every single letter of every word of the ad to get the gist. In the same way, a listener—especially a young one—doesn't need to be able to understand every motive of the characters or even know what all the words mean to become engrossed in a story. The speedwriting ad is not so very different in this respect from, say, "Jabberwocky," the nonsense poem from Lewis Carroll's second Alice book, *Through the Looking-Glass*. The verses famously begin and end:

> *'Twas brillig, and the slithy toves*
> *Did gyre and gimble in the wabe:*
> *All mimsy were the borogoves,*
> *And the mome raths outgrabe.*

Maria Tatar, who teaches Carroll to her students at Harvard, says, "You hear it, and it's a story, but you only understand half the words in it and it doesn't matter. It holds together. I often think

that must be the experience of a young child hearing stories—
getting half of it but still getting it."

The gap between receptive and expressive vocabulary—that
we understand more than we can say—is in some respects a mechanical issue of grammar, syntax, and context. In a wider sense,
it becomes a space in which we can experience the transcendent
and even mystical. Lewis Carroll can make up screwball words
and string them into a poem, and somehow we are able to supply
a meaning. The artistry of words can stir and excite us without
our knowing quite why. That was the experience of the novelist
Philip Pullman when he was a teenager and his English teacher
introduced the class to John Milton's epic poem *Paradise Lost*.

"Many of the references made little or no sense to me," Pullman told the children's book historian Leonard Marcus. "Nev-
ertheless, the sound of Milton's poetry when heard read aloud,
and then tasted afterward in your own mouth, was enormously
powerful. From that experience, I learned that things can affect
us before we understand them, and at a deeper level than we can
actually reach with our understanding. I also learned that you respond physically to poetry. Your hair stands on end. Your skin bris-
tles. Your heart goes faster."

I've seen this happen with my own children. I have seen them
respond to powerful writing in their gestures and breathing; there
is no doubt but that their hearts beat faster. The book that has
done it for them, every time, is *Treasure Island*. I have mentioned
that we read it every few years. This explains in part, I think, why
it has so captivated them. Each time we've read it, the children
were a step older than they'd been during the previous go-round.
It was clear to me that with each reading, they understood more
and more of the language and story, in a kind of time-lapse demonstration of the "again, again" phenomenon.

The first time I read it out loud, Molly was six and would have
sat close beside me, stopping me now and then to ask about an

unusual word or concept. She was good about that. Paris, at four, would have been playing with toys on the floor and following the story in a hazy way. Violet was less than a year old at the time, so I assume that all she heard was the reassuring blur-blur-blur of her mother's voice.

Two years later, we returned to the book, and this time Paris reacted to it as if he'd been shot out of a cannon. He was thrilled by the story, so electrified that he kept leaping up to act out the scenes. By the time we got to chapter twenty-five, I couldn't go two lines without an explosion. It is a very tense chapter. The young hero, Jim Hawkins, is in a tight spot. All alone, he's facing the wily, wounded coxswain, Israel Hands, who has lately been exposed as a pirate. The two confront one another on the deck of the *Hispaniola*, which has run aground just off Treasure Island, and as the surf knocks the hull, two dead buccaneers (one in a red cap) slide back and forth on the deck.

"'There were the two watchmen,'" I read, "'sure enough, red-cap on his back, as stiff as a handspike, with his arms stretched out like those of a crucifix and his teeth showing through his open lips; Israel Hands propped—'"

Paris vaulted off the sofa. He threw himself onto the floor, his arms rigid and straight out, and bared his teeth.

"Like this?"

We all laughed. Paris cannonballed back into his place.

"'—Israel Hands propped against the bulwarks, his chin on his chest, his hands lying open before him on the deck, his face as white, under its tan, as a tallow candle—'"

"What's tallow?" Molly asked.

"Beef fat."

"Bleah." She made a face.

"Go on!" Paris cried. "I love this story!"

"'For a while the ship kept bucking and sidling like a vicious horse, the sails filling, now on one tack, now on another, and the boom swinging to and fro till the mast groaned aloud under the strain—'"

"Psssshhhwww!"

Paris blasted off again, reeling around the room making a noise like a schooner in bad weather, his arms billowing like sails and whacking back and forth like a boom.

"Like this, right? Gnnnarrrr. . . ."

There was no mystery to his feelings about the book, no subtle, veiled inner magic or floating above the story, looking down. He was up to his gunwales in the adventure. It was unbelievably loud.

Would a boy his age have loved the story as much if this were the first time he was hearing it? Maybe. I think not, though: I think that it resonated with such power because somewhere in the back of his mind was an awareness that he'd heard it before when he was too young to grasp its subtleties. Coming to it a second time, Paris already had a feeling of ownership, a preexisting stake in the adventure. Like Alice tumbling down the rabbit hole, he was subsumed, surrounded, and transported.

Paris was eleven the next time *Treasure Island* came around. This time he didn't leap out of his seat. He had a new and more mature appreciation of the moral predicaments facing Jim Hawkins and his companions. In the charming villain Long John Silver, he saw the difficulty of distinguishing between a man's apparent nature and the one he really possesses.

Had Paris seen a film version of the story every couple of years, I can't imagine that he would have had such a profound and deepening series of encounters. As he listened, he had to draw from an inner well to conjure another world, with its smells and sounds and people: the metallic concussion of a spade on dry soil, the cold shock of seawater, the pink, shining, hamlike face of Long John Silver. A child watching a movie doesn't need to do any of that.

Film is a fabulous art form, but it does have a certain totalitarian aspect: the authorities have decided on the look and feel of everything from the clothes of the characters to the slant of the light to the mood and the music; the viewer's input is superfluous. It is for that reason that some parents try to keep ahead of Disney, the BBC, Walden Media, and other cinematic interpreters of classic children's literature. We want our children to taste the full flavor of *Peter Pan* and *Winnie-the-Pooh* and *Charlotte's Web* in the original, for themselves, before they see the big-screen renditions.

It was hard to keep ahead of Hollywood when my children were little, and with the Internet's role in family life it's harder for parents today. A first-grader who's not ready yet to tackle the Harry Potter books may not be able to avoid seeing the movies, or at least pictures of the actors. Before he's read a word of J. K. Rowling, in his mind Daniel Radcliffe is Harry, and Maggie Smith will be Professor McGonagall, forever. A story is never wholly yours again, once a movie has colonized your mind's eye. (Right before the new movie version of *A Wrinkle in Time* hit theaters in 2018, I rushed to read the book to Flora. It was too late: she had seen the promos, and from her questions, it was clear that she was "seeing" Madeleine L'Engle's novel through a lens crafted by the film's director, Ava DuVernay.)

*   *   *

IN ONE RESPECT, though, movies and reading aloud—and audiobooks—do have something important in common. They are sometimes accused of being easy shortcuts through the work of appreciating literature. There is some small truth to this, I suppose. When we sit still and watch a movie, it is true that we are passive rather than active. We are in receiving mode. It takes some modest effort to read Louis Sachar's *Holes*, whereas it takes only the parking of one's backside on a comfortable surface to enjoy the story on film.

A person listening to a novel read out loud might appear similarly passive. But it is not so: a story read aloud doesn't take place on the page, or even in the voice of the reader, but in an esoteric mingling of these things with the mind and the heart of the hearer. To listen to a story is to participate in its realization. Far from being a cheap shortcut, it's a deep and thoughtful way to engage with writing of all types.

*   *   *

WHEN WE READ aloud, we introduce small children to books as artifacts, as curiosities, as vessels of wonder and knowledge. They learn how books work: how the pages turn, how meaning flows from left to right, what letters look like and how they can be joined up into words, and words into sentences, and sentences into paragraphs.

As time goes on, and children get older, the reading will acquaint them with richer, more varied, more formal types of prose. They will learn to understand figurative devices ("stiff as a handspike," "a face like tallow") and to spot puns and rhymes and alliteration. They will know first-person from third-person narration. They will be able to tell the difference between stories told in the present tense and those in the past tense, and they will be familiar with dialogue and dialect. All this discernment will strengthen the cause of independent reading.

Is it cheating, though? Do we undermine young people, even infantilize them, by reading aloud long past the point that they can read for themselves?

No, it isn't. And no, we don't. Being read to and reading to oneself are, at heart, two ways of taking in one text. The mechanisms at the neurological level are different. But as an experience the distinction is a bit like the difference between walking and running. Both are good ways to reach a destination. Is it babyish to walk, which takes longer but requires less effort? Is it more mature to run, expending more energy and arriving sooner? There may be various things to consider when you choose between a run and a walk, but maturity doesn't enter into it. The same is true with reading in its silent and spoken forms. Depending on how we encounter words, our brains perform different functions to make sense of them. "We were never born to read," as Maryanne Wolf says, but must learn how. Yet our brains seem not to keep close records of how a story makes its way in. Matthew Rubery, the audiobook historian, told me that people who alternate between reading a physical book and listening to an unabridged audio version typically can't remember which passages they read using one method and which they "read" with the other.

And for people who struggle with reading, what a relief to be freed of the obligation to untangle sentences on their own! For them, hearing literature brought to life by the voice may be their single chance to have a meaningful, nonscreen encounter with Mr. Tumnus or Scheherazade or the Cheshire Cat. It may be the best means of enjoying novels with the ease and totality of their more adept and literate peers.

At school, adults can demand that young people read novels, test them on their comprehension, make them write reports, and expect them to be able to identify themes and motifs. It's not the worst thing in the world, but for many young people it's not a system that helps them fall in love with literature, either. Reading novels out loud offers another way in—in the process keeping listeners of all ages connected to books and to long-form narrative when their interest may otherwise have moved to screens.

In a 2016 essay, a middle-school teacher in Wisconsin named Timothy Dolan wrote that, "Each year I have a few students that come to loathe reading and writing. Many times it is due to low reading skills, but not always. They've spent so many years simply trying to figure out how to say the words in a text, instead of becoming immersed in the story itself."

Dolan's remedy is to read his students captivating, gritty novels such as *The Outsiders*, by S. E. Hinton, and Ray Bradbury's

Fahrenheit 451. This brings the material alive for the A students and the C-grade kids, in the same way and at the same time. The effect on the class is both unifying and equalizing. Every student is on the same footing as the story rolls out. "My low readers spend a lot of time deciphering words that are sight words for most grade-level readers," Dolan wrote. "How can they appreciate the beauty of Ray Bradbury's writing when so many words are painful obstacles? Reading aloud gives them the opportunity to hear complex texts without the onus being placed on their shoulders."

As to that question of infantilizing—well, it comes up in the classroom, too. "This year I had a parent question me on the validity of reading aloud with teenagers," Dolan wrote. "Initially I was offended, but then I remembered back to my first year of teaching. I had wondered the same thing. I couldn't imagine sitting in front of thirty eighth-graders and with a book having them actually pay attention. But I'll tell you what, when Ponyboy reads Johnny's letter at the end of *The Outsiders*, you can hear a pin drop in my classroom."

In a recent twist, some teachers are starting to use podcasts and transcripts in a similar way. Students put on headsets and listen to the same program at the same time while following the words on their tablets so that they are getting simultaneous aural and visual versions. High-school teacher Michael Godsey tried the technique using the first season of *Serial*, the popular podcast hosted by Sarah Koenig. His students were riveted by the real-life story of a dead Baltimore teenager and the former boyfriend who may or may not have killed her. Writing about his classroom experience in the *Atlantic*, Godsey acknowledged the tension between making young people do the work of reading and letting them listen.

"While I felt guilty the students weren't reading very much during this unit," he decided, "their engagement with a relevant and timely story—their eagerness to ask questions, their intrinsic motivation to use critical thinking—seemed to make it worth it, at least temporarily."

Godsey noticed that his students were much more engaged than usual. They argued with one another, consulted maps, wrote at length in their journals, and seemed eager to talk with adults about what they thought.

> Many of them said that reading along with the audio helped with their focus and kept them from "spacing out" while listening. Others, paradoxically, wrote that they were able to multi-task—they could take notes or write on their worksheets and could keep up with the story even with their eyes off the screen. Some explicitly recognized that they could look back and re-read something they didn't understand when they first heard it; others said they read slightly ahead and then could write down a quote while they listened to it. A student with eyesight problems said he appreciates the ability to take reading breaks without stopping his enjoyment of the story. A few students learning English as a second language wrote that they like how they can read the words and—as one student put it—promptly "hear how they're supposed to sound."

The headphones-plus-tablet approach may not have quite the same human warmth as a story read by a single voice to everyone at once. It's more like parallel play, with each student alone and separate. Still, it seems worthwhile. The teacher's selection of trendy, topical *Serial* is also a good reminder that to keep teenagers connected to oral storytelling may require some creativity in the choice of material.

That's true for older teens and adults, too, as Jane Fidler discovered when she took a job teaching remedial English at a Maryland community college. Her classes were full of people who had slipped through the cracks of the public education system. Many of her students were working full-time as well as trying to get a degree. A few of them were combat veterans. A lot of them struggled. One of Fidler's students, a young man who attended class through a day-release program, once came back from spring break without having done his homework because the prison where he lived had been on lockdown.

Most of her kids had never read a book all the way through before they got to community college, Fidler told me. "I say, 'How did you pass high school?' They say, 'I just wrote papers on books without reading them.'"

To get her students interested in fiction, Fidler decided to read them the juiciest, most accessible material she could find: salacious thrillers with short chapters and lots of action.

"In my lower developmental class, we read *Sail*, by James Patterson [and Howard Roughan]," Fidler said. "It's a very sexy book about a woman whose husband cheats on her, and she remarries, and he wants to kill her, and my students *love* it.

"'Okay,' I'll say, 'take out *Sail* and we're going to read chapter twenty-five.' And I'll read to them about how Peter Carlyle is two-timing his wife and playing around with his student, Bailey. One student said to me, 'I was up until four in the morning, finishing this book where you left off!'"

Fidler uses her unorthodox textbooks to teach specific lessons. She gets her students to draw inferences about what's going to happen to the characters. She explains vocabulary words. "I can help them focus on things that they would not have thought about if they read it on their own. And what I see happening is, when students get to the end of the book, they'll turn the page and see, 'Ooh, there's another book by Patterson, and this one sounds interesting.' This from kids for whom it is the first book they've read! That's pretty good."

It *is* pretty good. Knowing what I do now, I wouldn't hesitate to read a school assignment out loud if one of my children were having trouble with it. I wish we could go back in time and recoup the miserable hours that Phoebe spent wrestling with *Johnny Tremain* in the summer before fifth grade. It seems obvious to me now that the language and ideas were pitched a bit too far ahead of her. She couldn't read the novel with ease, so she couldn't read it with enjoyment. If I had brought it to life for her by reading it aloud, she might have relished spending time in revolutionary Boston with poor mangled Johnny and the rebellious Sons of Liberty. Rather than grinding through an ordeal that left her hating the author, Esther Forbes, she might have been able to appreciate the book's force and sentiment and beauty.

And isn't that the point? I mean, what else is the purpose of reading literature? Novels aren't supposed to be siege machines, or thumbscrews. In the long history of the humanities, there have no doubt been writers who set out to daunt the people reading their books, but most, surely, have had happier and higher ambitions.

*   *   *

SHORTLY BEFORE ROALD DAHL came into the world in the autumn of 1916, he was subjected to an eccentric prenatal educational scheme devised by his father.

"Every time my mother became pregnant," Dahl recounts in his memoir, *Boy,* "he would wait until the last three months of her pregnancy, and then he would announce to her that 'the glorious walks' must begin. These glorious walks consisted of him taking her to places of great beauty in the countryside and walking with her for about an hour each day so that she could absorb the splendor of the surroundings. His theory was that if the eye of a pregnant woman was constantly observing the beauty of nature, this beauty would somehow become transmitted to the mind of the unborn baby within her womb and that baby would grow up to be a lover of beautiful things."

The technique seems to have worked with Dahl, for he delighted in beautiful people, things, and ideas, even if, in his writing, he seems to honor beauty more in the breach than in the observance. Dahl's books are full of grotesques, monsters of appetite and vanity in the shape of greedy eaters, selfish parents, tyrannical headmistresses, and hideous, wig-wearing witches. In *Charlie and the Chocolate Factory*, *The BFG*, and *Fantastic Mr. Fox*—and in *Tales of the Unexpected*, a collection of stories for adults—his writing is full of pepper and zip. It never meanders, but gallops. Big things happen in his stories for children, and they happen with terrific force. Dahl's heroes and heroines are kind and imaginative. In the end, wickedness always goes down in defeat, and the shy and honorable prevail and flourish.

In reading Dahl's books out loud—and it's enormous fun to do so—we ourselves are, in a curious way, pointing our listeners toward beauty. This beauty is not the ethereal, sublime majesty of countryside you might traverse on a glorious walk. It is beauty of a puckish, radiant, comical, humane kind: the beauty of good winning out over evil, and the beautiful satisfaction of villains getting a thumping, rendered in writing so crisp and clean it seems to crackle as you read it. Consider the delicious moment in *James and the Giant Peach* when the colossal fruit detaches itself from its tree. The great sphere rolls toward James's two greedy aunts, Spiker and Sponge, who have expected to become millionaires from exhibiting the peach to the paying public.

To get the full flavor, try reading this aloud:

They gaped. They screamed. They started to run. They panicked. They both got in each other's way. They began pushing and jostling, and each one of them was thinking only about saving herself. Aunt Sponge, the fat one, tripped over a box that she'd brought along to keep the money in, and fell flat on her face. Aunt Spiker immediately tripped over Aunt Sponge and came down on top of her. They both lay on the ground, fighting and clawing and yelling and struggling frantically to get up again, but before they could do this, the mighty peach was upon them.

There was a crunch.

And then there was silence.

The peach rolled on. And behind it, Aunt Sponge and Aunt Spiker lay ironed out upon the grass as flat and thin and lifeless as a couple of paper dolls cut out of a picture book.

Notice how the scene thrums with action and sensation from the onomatopoeia of jostling, clawing, crunch, and silence. The giant peach is smooth and beautiful and as big as a house, its skin, Dahl writes, "a rich buttery yellow with patches of brilliant pink and red." How satisfying that an object so lush and gorgeous should be the means of obliterating a small boy's persecutors.

"Literature for children enthralls and entrances in large part through the shock effects of beauty and horror," Maria Tatar writes in her book *Enchanted Hunters*. When we spoke, she expanded on the idea. "We all like to be shocked and startled," she told me, "and there's something about amplification and exaggeration that you always get in these stories, and then you also get riddles and enigmas. You're shocked and startled, and curious—how did this thing happen? What if something like this happened? So immediately all your senses are engaged."

Beauty and horror work, in stories, to expose the many dualities of human nature: truthfulness and deceit, tenderness and hostility, loyalty and betrayal, generosity and greed. In Roald Dahl's work, the gruesome and freakish provide the reader with giddy delight even as their example illuminates the true and the good. In the Harry Potter books, J. K. Rowling explicitly links love and courage to sacrifice and loss; death hovers over the series like the Dark Mark writhing in the night sky.

Fairy tales, too, abound with monstrousness that has the effect of pointing to nobility and loveliness. Think of the jealous queen who sends a huntsman to cut out the heart of Snow White. Think, too, of his yielding in the face of the girl's beauty and guiltlessness, how he lets her slip away and returns to his mistress with the warm heart of a fresh-killed deer. As Vigen Guroian writes in *Tending the Heart of Virtue*, "By portraying wonderful and frightening worlds in which ugly beasts are transformed into princes and evil persons are turned to stones and good persons back to flesh, fairy tales remind us of moral truths whose ultimate claims to normativity and permanence we would not think of questioning. Love freely given is better than obedience that is coerced. Courage that rescues the innocent is noble, whereas cowardice that betrays others for self-gain or self-preservation is worthy only of disdain. Fairy tales say plainly that virtue and vice are opposites and not just a matter of degree. They show us that the virtues fit into character and complete our world in the same way that goodness naturally fills all things."

In the most powerful works of children's literature, danger and death are seldom far from loveliness, whether we've gone to Narnia or Oz or Terabithia. As novelist Jacqueline Woodson remarked when, in 2018, she became the US children's laureate, "When we come out of a book, we're different."

*   *   *

THE ENCHANTMENT OF a piece of writing delivered by the human voice may come on little cat feet, so to speak, slipping in so softly that we hardly notice its arrival. It can also dash forward and strike with a blow, as it did one night in 1917 to the future novelist and folklorist Zora Neale Hurston. Hurston had enrolled in a night high-school English class taught by a man named Dwight O. W. Holmes.

In her memoir, *Dust Tracks on a Road*, Hurston wrote, "There is no more dynamic teacher anywhere under any skin. . . . He is not a pretty man, but he has the face of a scholar, not dry and set like, but fire flashes from his deep-set eyes. His high-bridged, but sort of bent nose over his thin-lipped mouth—well, the whole thing reminds you of some Roman like Cicero, Caesar or Virgil in tan skin."

One fateful evening, this teacher opened a volume of English poetry and began to read to the class:

> *In Xanadu did Kubla Khan*
> *A stately pleasure-dome decree:*
> *Where Alph, the sacred river, ran*
> *Through caverns measureless to man*
> *Down to a sunless sea.*
> *So twice five miles of fertile ground*
> *With walls and towers were girdled round;*
> *And there were gardens bright with sinuous rills,*
> *Where blossomed many an incense-bearing tree;*
> *And here were forests ancient as the hills,*
> *Enfolding sunny spots of greenery . . .*

Hurston was transfixed. "Listening to Samuel Taylor Coleridge's *Kubla Khan* for the first time, I saw all that the poet had meant for me to see with him, and infinite cosmic things besides. I was not of the work-a-day world for days after Mr. Holmes's voice had ceased. This was my world, I said to myself, and I shall be in it, and surrounded by it, if it is the last thing I do on God's green dirt-ball."

What happened to Hurston that night was a kind of intellectual and aesthetic liberation: the sound of her teacher reading Coleridge was so thrilling, so arresting, that it cut her loose from the life she might have had and freed her to find her destiny as a writer.

At other times in history, and in other places, reading aloud has been the means of more literal liberation. The human voice is just a sound in the air, and yet it has built bridges from ignorance to knowledge, and from bondage to freedom. In the American South before the Civil War, for instance, it was illegal in some states to teach enslaved people to read and write. There were no laws against listening, though, which is how the future abolitionist and writer Frederick Douglass got his first taste of what words could do. He was about twelve at the time, and as he later wrote, "the frequent hearing of my mistress reading the Bible aloud awakened my curiosity in respect to this mystery of reading, and roused in me the desire to learn."

The woman began to teach Douglass the letters of the alphabet, but her husband soon put a stop to it. After that, Douglass recalled, her attitude toward him changed. "Slavery proved as injurious to her as it did to me," he wrote. "Under its influence, the tender heart became stone, and the lamb-like disposition gave way to one of tiger-like fierceness."

At one point, Douglass's mistress became so enraged at the sight of him holding a newspaper that she yanked it from his hand. "She was an apt woman," Douglass observed dryly, "and a little experience soon demonstrated, to her satisfaction, that education and slavery were incompatible with each other."

Like Douglass, the future missionary and preacher Thomas Johnson paid close attention to readings of the New Testament at night. Johnson would ask to hear certain passages repeated, so that he could fix the words in his mind, and then he would compare what he'd heard with what he saw printed in a stolen Bible that he kept hidden away. Reading aloud became, for these determined men, a secret staircase that led to the open air of intellectual escape.

When a cell door slammed on Yevgenia Ginzburg, a Communist Party official caught in the purges of Stalin's Great Terror, she was left with one source of consolation: "Poetry, at least, they could not take away from me!" she declares in her memoir, *Journey into the Whirlwind*. The prisoner prowled her cell, racking her memory to recite aloud the literature she'd read. "They had taken my dress, my shoes, my stockings, and my comb . . . but this was not in their power to take away, it was and remained mine."

A few years later, eight hundred miles to the west, retelling literature from memory came to the rescue for Helen Fagin, a young prisoner of the Warsaw ghetto. "Being caught reading anything forbidden by the Nazis meant, at best, hard labor; at worst, death," she writes in an essay for the collection *A Velocity of Being*.

> I conducted a clandestine school offering Jewish children a chance at the essential education denied them by their captors. But I soon came to feel that teaching these young sensitive souls Latin and mathematics was cheating them of something far more essential—what they needed wasn't dry information but hope, the kind that comes from being transported into a dream-world of possibility.
>
> One day, as if guessing my thoughts, one girl beseeched me: "Could you tell us a book, please?"

Fagin had spent the previous night devouring a contraband copy of Margaret Mitchell's *Gone with the Wind*, so her own dream-world was still "illuminated" by the story.

> As I "told" them the book, they shared the loves and trials of Rhett Butler and Scarlett O'Hara, of Ashley and Melanie Wilkes. For that magical hour, we had escaped into a world not of murder but of manners and hospitality. All the children's faces had grown animated with new vitality.
>
> A knock on the door shattered our shared dream-world. As the class silently exited, a pale green-eyed girl turned to me with a tearful smile: "Thank you so very much for this journey into another world. Could we please do it again, soon?" I promised we would, although I doubted we would have many more chances.

Only a few of the children in the secret school survived the Holocaust. The green-eyed girl was one of them. "There are times when dreams sustain us more than facts," Fagin concludes. "To read a book and surrender to a story is to keep our very humanity alive."

*   *   *

IT IS NO accident that repressive governments often limit people's access to books and information. That was true in Spanish Cuba, when the authorities put a stop to the public readings in the cigar factories. That was true in the Warsaw ghetto for Helen Fagin. Books consumed in private cultivate independence of mind, a thing unwelcome and even dangerous when the culture outside is in the grip of orthodoxy.

The experience of Chen Guangcheng, the blind human rights activist who made a dramatic escape from house arrest in China to the American embassy in 2012, speaks to the power of reading out loud as a means not only of liberating the listener's imagination but also of engaging his critical faculties. If those faculties happen to be subversive, well, the responsibility for that doesn't lie with the thinker but with those who would stop him from thinking.

Chen was born in 1971, and in rural China the blindness that took his sight after a fever meant that he could not get a formal education. As a result, he spent his days isolated from other children. While the rest of the kids in his village attended the Communist Party–run local school, Chen spent his time trapping frogs, devising gimcrack homemade guns, and building kites that he couldn't see, but whose airborne vibrations he could feel through the string in his hand. His mother was illiterate, but his father had picked up the rudiments of reading and writing just before the Cultural Revolution shuttered the schools in 1966. In the horror and tumult of the next decade, young Red Guards plundered libraries and ransacked temples, smashing and burning books and antiquities.

They brutalized and rusticated the educated, the once-prosperous, and the insufficiently zealous in a state-sanctioned campaign to extirpate the "four olds": old ideas, old customs, old habits, and old culture.

The fever would pass, but even as Chinese society was shuddering in the aftermath of the revolution, Chen Guangcheng's father was doing something extraordinary. Quietly, every night, he read to his blind son. In doing so, he imparted old ideas, old customs, old habits, and old culture.

"My father and I would sit under the kerosene lamp as he read aloud, making out the words in a halting rhythm, his voice rough and low," Chen recalls in his memoir, *The Barefoot Lawyer*. Their books ran from folktales to history to Chinese classics. Father and son read the sixteenth-century novel *Investiture of the Gods*. They read the sprawling, tragic eighteenth-century love story *Dream of the Red Chamber* (banned during the Cultural Revolution). They read the fourteenth-century epic *Romance of the Three Kingdoms*, also banned, on the grounds that it encouraged mythology.

Hour after hour, sometimes sitting up, sometimes lying by his father's side on a narrow bed, the boy listened. "The stories my father read to me served as a counterpoint to the official party line and the usual propaganda," Chen writes.

Just as important was that my father's stories and our discussions about them gave me an organic education in ethics, providing a framework with which to understand my experience as a disabled child. The stories I heard when I was young allowed me to imagine myself in the position of the characters, to consider how I would react if faced with similar challenges, to devise my own responses and then to compare them with what actually took place.

Chinese history is full of examples of the disempowered overcoming the odds through wit and daring. Though I lacked

the conventional education of my peers, I also avoided the propaganda that was part and parcel of the party's educational system. Instead, my father's tales became my foundational texts in
everything from morality to history and literature and provided
me with a road map for everyday life.

It is a brilliant testament to the dedication of a loving father,
and to the excellence of *five* olds: old ideas, old customs, old habits,
old culture, and the old practice of reading aloud.

Wherever young people are growing up, they deserve to know
what went into the making of their world. They have a right to
be free to enjoy the richness that history and culture have bequeathed them.

By reading aloud, we can help make that happen.

CHAPTER 7

READING ALOUD
FURNISHES THE MIND

———◦•◦———

Young children, as we clearly see,
Pretty girls, especially,
Innocent of all life's dangers,
Shouldn't stop and chat with strangers.
If this simple advice beats them,
It's no surprise if a wolf eats them.

—Charles Perrault, *Tales of Mother Goose*

We almost never take this out because it is really fragile," said Christine Nelson, a curator at the Morgan Library in New York. I was sitting across from her, in her office. She drew out a small navy-blue case and opened the lid. Inside, its glossy red leather binding embossed with gold, was the earliest surviving volume of the fairy tales of Charles Perrault. This beautiful object had been created in 1695 as a gift for the teenage niece of Louis the Fourteenth, a girl known as "Mademoiselle."

Nelson opened to the frontispiece, revealing a charming little painting. A plain-faced woman in a linen coif and rustic dress sits before a fire, holding a spindle of wool. She seems to be telling a story to three young people in elegant clothes, one of whom leans forward, touching the storyteller's knees in her eagerness. Curled up by the fire, a plump little cat listens, too. On the wooden door behind the spindle holder, a sign reads: "Contes de Ma Mere l'Oye."

Tales of My Mother Goose! More than three centuries ago, a careful hand (possibly Perrault's son, Pierre) had dipped a pen in ink and in beautiful cursive committed the world's first known collection of fairy tales to this folio. Now the pages were fragile, crisp, and speckled with age spots.

There I was, sitting in a modern office building, with trucks and cars rumbling up nearby Madison Avenue, and for a fraction of a second the book before me seemed to become a portal, like a wardrobe into Narnia or a portkey at Hogwarts, that could fling me into the past. I had the fleeting idea that if I were to touch the page, I might be flashed back to a place of silks and mirrors and a laughing girl, and that if I were to squint or tip my head at the right angle, I might go deeper still, through the story and out the other side, into the hazy Indo-European folkways where the stories began. It was the impression of a moment, and it was whimsical, I know, but the tales that Perrault collected have such broad cultural resonance today that I felt giddy to be so close to *the* first Mother Goose.

Charles Perrault is credited with creating the literary tradition of the fairy tale, but of course the stories he told weren't his. They had come from deep in the trackless past and were, by word of mouth, on their way into the future when he plucked them from the air and wrote them down. He and other collectors and folklorists over the centuries and across the world—enterprising individuals such as Marie-Catherine d'Aulnoy, Wilhelm and Jacob Grimm, Andrew Lang, Moltke Moe, Lafcadio Hearn, Charles Chesnutt, W. E. B. Du Bois, and many others—have preserved vast libraries derived from "the golden net-work of oral tradition" that might otherwise have been lost. Without their efforts, we would have inherited nothing like the richness of story, song, and legend available to us in the digital age.

I was thinking about this as Christine Nelson turned the page so that we could see Perrault's version of "Le Petit Chaperon rouge"—*Little Red Riding Hood*, a tale so old that scholars have traced some of its elements to classical Greece and the myth of the child-swallowing Titan, Cronos. According to Bruno Bettelheim, the story has been told in France in one form or another since at least the eleventh century; he mentions a tale from that period, recorded in Latin, that tells of a little girl wearing a red cover, or cap, who's found in the company of wolves.

In the Morgan Library's edition of Perrault, the story of *Little Red Riding Hood* begins with another perfect little gemlike painting. In this one, a woman reclining in bed has raised herself up and seems almost to be greeting a large dog that has bounded up. The animal's front paws are resting on the bed's crimson coverlet. His hindquarters are hidden beneath its golden canopy. He is no dog, of course, but a wicked wolf: the artist has captured the moment of the grandmother's destruction.

Underneath the picture, delicate French cursive writing tells the celebrated story. Just near the end, the scribe has inked an asterisk next to the famous lines of dialogue: "What big teeth you have!" and "The better to eat you with." Tiny handwriting in the margin instructs the reader: "On prononce ces mots d'une voix forte pour faire peur a l'enfant comme si le loup l'alloit manger" (These words should be said in a loud voice to make the child afraid that the wolf will eat him). We know from this that Perrault didn't create this exquisite collection for private enjoyment. He expected Mademoiselle to read the stories out loud.

*   *   *

"IF YOU WANT your children to be intelligent, read them fairy tales. If you want them to be more intelligent, read them more fairy-tales," Albert Einstein advised. I don't know if the great theoretical physicist really made that remark, and I cannot promise that reading fairy tales to a child will tweak his IQ, but there is no doubt that these weird dramas of risk, terror, loyalty, and reward agitate the blood and captivate the heart. To C. S. Lewis, time spent in what he called "fairyland" arouses in a child "a longing for he

knows not what. It stirs and troubles him (to his lifelong enrichment) with the dim sense of something beyond his actual reach
and, far from dulling or emptying the actual world, gives it a new
dimension of depth. He does not despise real woods because he
has read of enchanted woods: The reading makes all real woods a
little enchanted."

The reading does something else, too. It situates children in a
cultural sense, equipping them to understand references to fairy
tales and other classic stories that they will find all around them.
When we read *Hansel and Gretel* or *The Fisherman and His Wife* or
Puss in Boots, we're at once transporting children with our voices
and grounding them in foundational texts. For this reason, the
time we spend reading to them can amount to a second education, one that helps children "acquire a sense of horizons," in the
phrase of linguist John McWhorter. What we give them is not
schooling *qua* schooling, but an introduction to art and literature by means so calm and seamless that they may not notice it's
happening.

We can furnish their minds with eccentric oddments, beautiful images, and useful bits of general knowledge. We can introduce them to a world of peculiar and renowned characters: to Ali
Baba and the Forty Thieves, to Custard the Dragon, to Corduroy,
Strega Nona, and the Farmer in the Dell; to the Sheriff of Not-
tingham and the Velveteen Rabbit; to Baba Yaga and her house on
chicken legs; to the boy who cried wolf, to Apollo and Artemis,
to Daedalus and Icarus; to Coraline, Despereaux, the Runaway
Bunny, and the tricksters Loki and Anansi and the Cat in the Hat.

We can give them what, in truth, is already theirs. The books
and artwork of the world are, after all, the inherited property of
every child. They are the natural estate of every boy or girl from
the moment that he or she draws breath. Nursery rhymes, fairy
tales, legends, myths, poetry, paintings, sculptures, the great body
of classic literature, the bounty of new and forthcoming litera-

ture, whether for children or adults—all these things belong to the young and ignorant just as much as they do to the old and erudite. When the Dominican-born writer Junot Díaz was a boy, his elementary-school librarian in the United States showed him around the stacks and told him, he said years later, that "all the books on the shelves were mine." It was a galvanizing moment that he never forgot. It is a message that every child needs to hear.

Owning something and taking possession of it are two different things, of course. A child may have as much right to *Beowulf* as the scholar who devotes his life to the study of Old English. Yet unless the child meets the hero Beowulf and the monster Grendel, and Grendel's gruesome mother, he cannot be said to have taken possession of the property that's his. Let that child's mother read *Beowulf* in translation at bedtime (if she dares—it's pretty gory), and the child's custody is complete. The characters and scenes and language of the book will become part of his interior landscape. The book's mystical qualities will add sublimity to his experience of life.

The more stories children hear, and the more varied and substantial those tales, the greater the confidence of their cultural ownership. They will recognize allusions that other children may miss. A girl who has heard the stories of Aesop or Jean de la Fontaine will have a clear idea of what is meant by "sour grapes" and will know why people compare the industriousness of ants and grasshoppers. A boy who's heard a parent read *The Odyssey* has a more complete idea of what constitutes a "siren song" than his friend who thinks it must have something to do with an alarm going off.

The narratives of the past have helped to frame the consciousness and language of the present, and it's a gift to children to help them recognize as much as they can. The milk of human kindness, the prick of the spindle, the wolf in sheep's clothing, the wine-dark sea: all are expressions of a vast cultural treasury.

"We all come from the past, and children ought to know what it was that went into their making, to know that life is a braided cord of humanity stretching up from time long gone, and that it cannot be defined by the span of a single journey from diaper to shroud," Russell Baker writes in his beautiful memoir, *Growing Up*.

Children get a wider perspective when they're tugged out of the here and now for a little while each day. In an enchanted hour, we can read them stories of the real and imagined past. With picture-book biographies we can acquaint them with people we want them to know: Josephine Baker and Amelia Earhart, Julius Caesar and Marco Polo, Martin Luther King Jr. and Wolfgang Amadeus Mozart, George Patton and Shaka Zulu, Pocahontas, Frida Kahlo, Edward Hopper, William Shackleton, the Savage of Aveyron, and the terrible Tudors.

With any luck, our children will come to appreciate that the people of generations past were as full of life, intelligence, wisdom, and promise as they are, and impelled by the same half-understood desires and impulses; that those departed souls were as good and bad and indifferent as people who walk the earth today. Those who came before us wrote stories and songs, built roads and bridges, invented and created and argued and fought and sacrificed for all sorts of causes. Do we not owe them a debt of gratitude? We wouldn't be here without them.

Young people are inclined to think, in a vague way, that events began when they did. When I was a child, I was told that President Kennedy was shot in 1963. It seemed to me that the tragedy coincided, more or less, with the end of the Civil War. When you're young, the decades blur together. Only years later did I come to understand that JFK died six months before I was born, and that I entered a world still shocked by his departure.

So it goes. Youth is inattentive. It thinks itself something fresh, full of energy, spirit, and insight. It feels that no one has ever cared so much, felt with such intensity, or realized truth with such exquisite clarity. It prepares for a future that is unique in its grandeur and meaning. Youth may have no idea that it is wreathed in ghosts, informed by ghosts, held up on the shoulders of ghosts. When we read aloud from the literature of the past—and all literature is the literature of the past—and when we share artistic traditions, we are not merely giving children stories and pictures to enjoy. We're also inviting a measure of humility, gently correcting youth's eternal temptation to arrogance.

*   *   *

"TRADITION IS TO the community what memory is to the individual," said the Irish poet and Celtic mystic John O'Donohue. "And if you lose your memory, you wake up in the morning, you don't know where you are, who you are, what ground you're standing on. And if you lose your tradition, it's the same thing."

But what tradition? Well, it depends. Cultural literacy is a complicated concept, and what goes into it will depend on where and when a person lives. Today a modern American child's informal education still begins with nursery rhymes and songs. Many of the best-known ditties and verses date back centuries, and have disputed and murky origins in the commerce and politics of the British Isles. Their durability at this point may seem a little mysterious—who is Solomon Grundy to us, that we should care for him?—and yet they matter, because, as cultural signifiers, they're embedded all over the place.

To take an example dear to the heart, *Goodnight Moon* assumes a not-insignificant measure of background knowledge. You will remember that the paintings in the great green room show first "three little bears sitting on chairs," which refers to *Goldilocks and the Three Bears*, and, second, "a cow jumping over the moon," which alludes to the nursery rhyme "Hey, Diddle, Diddle."

The recognition of these foundational songs can go both ways. A child who encounters *Goodnight Moon* without knowing a nursery rhyme might come later to a Mother Goose collection and think, Say, haven't I seen a cow jumping over the moon somewhere else? But if he only sees *Goodnight Moon*, and never hears "Hey, Diddle, Diddle," the cow jumping over the moon will be an untethered idea with no deeper resonance. The child who has met the cow in the song, and then sees a reference to it in *Goodnight Moon*, on the other hand, will identify even more firmly with the little rabbit in the great green room: Hey, he knows the same stories and rhymes that I do!

In this way, cultural allusions intertwine and self-reinforce, and as with new vocabulary words, the more of them children know, the more they'll catch as they fly by.

It helps that children love the bump and tumble of rhyming words. They may not mind one way or another about the characters of Humpty Dumpty or Little Miss Muffet or Old Mother Hubbard, but most little ones find the lolloping rhythms of nursery songs irresistible. (I have the faintest memories of my mother chanting "Ride a cock-horse to Banbury Cross" to me. Holding my hands, she'd jog me on her knees, as her own mother had done with her. At the crucial line, "And she shall have music where-*ever* she goes!" she'd pretend to drop me, and I'd shriek. This all came back to me with a rush when I saw her do the same with my daughter Molly, as she, Molly, will probably someday do with her own children.)

Fun to chant, and culturally grounding to learn, nursery rhymes also happen to be a terrific entry point to language. Neurobiologist Maryanne Wolf and others believe that exposure to these traditional poems can help to sharpen a small child's awareness of the smallest sounds in words, known as phonemes. "Tucked inside 'Hickory, dickory dock a mouse ran up the clock' and other rhymes," Wolf writes in *Proust and the Squid*, "can be found a host of potential aids to sound awareness—alliteration, assonance, rhyme, repetition. Alliterative and rhyming sounds teach the young ear that words can sound similar because they share a first or last sound."

The writer, illustrator, and read-aloud evangelist Mem Fox agrees. "Children get a real kick out of the bounce and wackiness of poetry," she remarks in *Reading Magic*. "Once children have masses of rhythmic gems in their heads, they'll have a huge store of information to bring to the task of learning to read, a nice fat bank of language: words, phrases, structures and grammar."

So yes, we start with nursery rhymes and read them again and again—maybe setting them to improvised tunes, maybe reading versions in Spanish or French or another language, for fun. Then come traditional folk and fairy tales.

Fairy tales are so resonant and strange, so fantastically generative, that all the stories and all the versions of the stories and all the subsequent reimaginings of the stories and all the scholarship *into* and exegesis *of* the stories could fill the Black Forest itself—or perhaps the Sahara, for in truth they arise in every corner of the world. To take one example, variants of *Beauty and the Beast*, with the courtship and pairing of a beautiful human and a charismatic animal or monster, can be found in Zulu and Native American cultures, in Bolivia, Burma, Iran, India, Russia, Japan, Ghana, and elsewhere.

Why do these stories stay with us, and why, like nursery rhymes, do they continue to matter? This mystery is part of the attraction of these ubiquitous and ancient works of art. They appall us with horror and entrance us with loveliness. They jar and surprise. The neutrality of their characters ("the third brother" or "the queen" or "Jack") allows anyone to occupy the roles in imagination.

Fairy tales also encode a measure of philosophical and practical wisdom, as Vigen Guroian observed. It is certainly easy to see the moral precepts in *Little Red Riding Hood*. Perrault all but spells them out with the ominous sexual overtones in his version of the tale. Having devoured the grandmother and taken her place in bed, the wolf tells Little Red, when she arrives, to undress and join him. She does—to her ruin. In case we missed the metaphor, Perrault makes it explicit in the second half of the moral of the story (the first was at the beginning of this chapter):

> *And this warning take, I beg:*
> *Not every wolf runs on four legs.*
> *The smooth tongue of a smooth-skinned creature*
> *May mask a rough and wolfish nature.*
> *These quiet types, for all their charm,*
> *Can be the cause of the worse harm.*

Viewed in this cautionary light, *Little Red Riding Hood* seems to come to us a from a long line of parents stretching back in time, all calling, one generation after another, "Watch out! Beware the sweet-talking stranger!"

Bruno Bettelheim, in *The Uses of Enchantment*, argues that fairy tales offer children access to the quiet wisdom of the ages: "These tales are the purveyors of deep insights that have sustained mankind through the long vicissitudes of its existence, a heritage that is not revealed in any other form as simply and directly, or as accessibly, to children."

The vicarious journeys that children take into cottages and castles and through dark woods, their encounters with heroes, gnomes, giants, cannibals, talking animals, wish-granting fairies, witches and wolves and huntsmen with terrible, shining knives—all this comes to them, Bettelheim says, as "wondrous because the child feels understood and appreciated deep down in his feelings, hopes, and anxieties, without these all having to be dragged up and investigated in the harsh light of a rationality that is still beyond him. Fairy tales enrich the child's life and give it an enchanted quality just because he does not quite know how the stories have worked their wonder on him."

Much the same happens with myths and legends. Orpheus and Ariadne and the Minotaur are more distinct as individuals than the featureless, elastic characters of fairy tales. Yet their struggles and quandaries reveal universal aspects of life and human nature that children will recognize, in that dim, half-hidden way. Reading aloud stories of the Greeks, Celts, Persians, Romans, or Vikings helps to lay an intellectual foundation that is made up of sense and feeling, reason and knowledge. Children experience the stories as emotional events, even as they're picking up and stowing away important cultural references.

Once begun, a child keeps acquiring. In this way, the language, stories, and pictures of a person's early years build an imaginative scaffold. With exposure to a wide variety of images and narrative, the scaffolding gets stronger and more substantial. In time, it becomes a kind of interior reference library fully furnished with peculiar and interesting things.

*   *   *

WHEN IN 1943 Antoine de Saint-Exupéry published his slim fable *The Little Prince*, one reviewer described its effect in a remarkable and prescient way. The book, wrote the critic, "will shine upon children with a sidewise gleam. It will strike them in some place that is not the mind and glow there until the time comes for them to comprehend it." The reviewer was P. L. Travers, the author of *Mary Poppins*, who knew a thing or two about symbolism and submerged emotion.

The phenomenon that Travers described is, in some respects, the source of the reverberating magic of the best children's literature. When we are adults, we can look back at the books we loved in childhood and can remember the illustrations or stories (or paragraphs in stories) that meant something to us, even if we can't quite recapitulate the intensity of our first response. But why it mattered at the time? That is the mystery. As the great Robert

Lawson once observed, "No one can possibly tell what tiny detail of a drawing or what seemingly trivial phrase in a story will be the spark that sets off a great flash in the mind of some child, a flash that will leave a glow there until the day he dies."

It's true. We can't know when the flash might happen, or what will strike a spark from the flint, but we can encourage this incendiary process by making deliberate, adventurous, and sophisticated choices in the books we share with children.

*   *   *

ONCE UPON A time, a child might have to wait until high school to encounter classics of adult literature and masterpieces of fine art. Inexpensive board books now make it possible to bring these titles and pictures into the nursery, to be chewed upon by babies and hopefully absorbed in other ways, too. Jennifer Adams's Baby-Lit books, illustrated by Alison Oliver, use the titles and characters of great novels as starting points to teach colors, categories, and other concepts. For the Cozy Classics series, Holman Wang and Jack Wang hew much closer to the original stories, dramatically condensing novels such as *War and Peace*, *The Adventures of Tom Sawyer*, and *Les Misérables*, and illustrating them with photographed tableaux of soft felted figurines. The hope is that children will develop such warm, positive associations with these classic titles that they'll want to pick up the unabridged novels when they're older.

I met Jack Wang at a café near Ithaca College in upstate New York, where he teaches English. "We think the problem is that these classics have become overly academic," he said, "and people only see them in overly studied terms, and they've become intimidated by them. They're afraid to enjoy them as stories, which is what they are, first and foremost, right? Really compelling stories that have lasted because they told a great narrative."

In the Cozy Classics, great narrative is condensed to a comical degree. The text of the Wang brothers' version of *Moby-Dick*, for instance, reads in full: Sailor, Boat, Captain, Leg, Mad, Sail, Find, Whale, Chase, Smash, Sink, Float. This leaves out a good deal—206,040 words, to be exact—but does convey the gist and some of the drama, if not the humor, of Melville's original.

"The thing is, these stories belong to everyone," Wang pointed out. "It's not just the Western canon, per se, but everyone can feel a sense of ownership over these classics because these are great human stories. I want my kids to be close to these stories because it's part of their cultural inheritance too."

* * *

CULTURE DOES NOT consist solely of art and writing, of course, but also of attitudes, practices, and values. The things we say when we're talking to children about stories and pictures, the emphases we make and the bits we skip over, tell them something about how *we* see the world. A wordless picture book gave researchers at New York University a fascinating look at the way parents can impart distinct cultural norms when they're reading with children. For the 2015 study, clinicians video-recorded Dominican, Mexican, and African American mothers sharing Mercer Mayer's book *Frog, Where Are You?* with their young children. Building on similar observations of white, Latino, and Chinese mothers, the clinicians could compare how parents of different backgrounds infused the story with their own priorities.

"We found that the mothers constructed different stories, and those stories aligned with their expectations, and their emphases on what's important to know and to learn," Catherine Tamis-LeMonda told me. "The Chinese moms tended, when they read books or told stories, to talk a lot about the moral lesson, the social rules, and what you should do or not do. They would say things like: 'Oh, the boy touched the beehive! He shouldn't have done that. They can come out and hurt him, you don't do that!' They were using the story as a way to enforce rules and obligations. The Latinos were using the book to talk more about emotions and how people felt. And the African Americans were using the book to talk about goals, and how hard the boy has to work to find the frog, and persist in finding the frog, which put an emphasis on individual persistence and working hard."

No child is an island, to paraphrase John Donne. Children come from families. They are the newest braids in that cord of humanity, and it is right and beautiful that they should know something of what their parents and grandparents value, while at the same time having access to the classic works of human imagination that we all own in common.

Contemporary culture will take care of itself. It's lively and loud, and most children's lives are full of it. Electronic media will keep them up-to-date. When parents read long-beloved classics with them, and share stories that help us convey what we want them to know about the world, we can help them discover powerful narratives and pictures they will never find on PBS Kids or Instagram.

*   *   *

"THE IMAGES OF things impress themselves in our minds," said the Renaissance humanist Leon Battista Alberti. And who can doubt it? What we look at determines what we see, and what we see—really *see*—becomes part of an inner museum of pictures and references, a mental collection that for most of us is not so much curated as acquired in a haphazard way.

There is an opportunity, with children, to show them art and illustration that will furnish their minds with beauty and mystery, symmetry and wonder. The simplest mechanism for this is the selection of picture books that we share with them. The beauty they find may be sweet and domestic, like the scene of little Peter's hot bath after his adventures in *The Snowy Day*, by Ezra Jack

Keats, or the rabbit father tucking his children into a stack of four bunk beds in Richard Scarry's illustrations for *The Bunny Book*, by Patsy Scarry. It may be elegant and dramatic, like Walter Crane's sumptuous princesses and Arthur Rackham's dreamlike ogres. These last come to us from a period known as the Golden Age of Illustration, which ran from the late nineteenth century into the early decades of the twentieth century and marked a kind of late-blooming, secondary renaissance involving artwork and graphic design for books and magazines. It is the era that brought us Kate Greenaway's graceful country girls, John Tenniel's sharp, alarming caricatures, and the rich, realistic paintings of Howard Pyle, Jessie Willcox Smith, and N. C. Wyeth, whose portraits of ruddy, ruthless pirates for *Treasure Island* will, for many devotees, forever represent the "gentlemen of fortune" described by Robert Louis Stevenson.

A little later, in the midcentury afterglow of the Golden Age, all manner of other extraordinary artists turned their talents to making pictures for children's books: Eloise Wilkin, with her round-cheeked innocents; Garth Williams, master of soft, evocative charcoal drawings; and Gustaf Tenggren, whose wild, wonky lines and colors brought to life the *Tawny, Scrawny Lion*, and many other characters. This was the heyday of the incomparable Margaret Wise Brown, who was herself not an illustrator but whose writing created opportunities for Clement Hurd, with his color-saturated pictures, and the blocky brilliance of Leonard Weisgard.

I mention these nineteenth- and twentieth-century illustrators because although the official Golden Age may have ended, their work is still with us. We and our children are fortunate: we have access to the best of the past and to the best that's being created today. You might say we are enjoying a new Golden Age, one informed as much by those earlier artists as by the scraggly energy of contemporary illustrators like Quentin Blake, Matthew Cordell, and Lauren Child; by the chilly genius of Chris Van Allsburg and

Jon Klassen; by the vibrant colors of Christian Robinson, Ana Castillo, and Raúl Colón; by the observant lines of Erin Stead, Brian Floca, and Barbara McClintock; and by the joyful watercolors of Suzy Lee, Jerry Pinkney, Chris Raschka, Meilo So, and Helen Oxenbury.

Beauty may be in the eye of the beholder, but the eye itself can be coaxed, informed, and persuaded. Like the mothers in the NYU study who each brought her own cultural understanding to a wordless book, we too can use the time we spend with children showing them what *we* think beautiful. We can also make it clear to them that fine paintings and sculpture—like poetry and novels—are not remote, forbidding objects that belong to the adult world, but sophisticated expressions of human creativity that belong to them, too.

"Making a connection to art is huge. When children have seen a painting at home, and then they see it in person, it becomes theirs," said Amy Guglielmo, a writer and artist who with Julie Appel created a board-book series called Touch the Art. Guglielmo was teaching kindergarten, using lots of artwork in her lessons, and took her class one day to the Museum of Modern Art in New York. Everything was fine until the children got to a particular gallery. "One of them recognized Picasso's *Three Musicians*," Guglielmo told me, "and he *touched* it."

There was a rebuke from the museum guards—never a nice experience—but the child's desire to make physical contact with a familiar picture was an inspiration. Guglielmo and Appel went on to create books featuring artwork from a wide range of periods and styles with each page, and each piece, having its own embedded tactile element. In the Touch the Art books, children can stroke the fur of Albrecht Dürer's hare, fiddle with the edge of a real tablecloth in Romare Bearden's *Autumn of the Rooster*, or tug the tutu of a Degas ballerina.

It is hard not to touch the art, to be honest, whatever your age.

A moment ago, my daughter Phoebe came in to my office and saw a copy of one of the books, *Brush Mona Lisa's Hair*, on my desk. "I love that one!" she cried, and, as she had when she was little, ran her fingers through a tangle of Mona Lisa's hair, stroked the yarn tail of a Velázquez horse, pinged the golden elastic around the hair of Botticelli's *Venus*, and toyed with the gilded lace standing out from the collar of a rakish fellow in a portrait by Frans Hals.

"Do you still want to marry him, Mom?" Phoebe smirked, with a teenager's annoying steel-trap memory. (Years ago, I had confessed to having a crush on *The Laughing Cavalier*.) She didn't wait for an answer, but flipped on, running her finger along the fur on the groom's robe in Jan van Eyck's *Arnolfini Portrait*, touching the felt hatband and raised diamond necklace of Piero della Francesca's double portrait of Federico da Montefeltro and his wife, Battista Sforza, tapping a pearl earring in the famous Vermeer portrait, and hunting for the pop-up window that reveals a secret ace of diamonds in Georges de la Tour's *The Card-Sharp*. She didn't bother with the final picture, a detail of Raphael's *Sistine Madonna*, because time and use had shorn the two cherubim of their fluffy, claret-colored feathers.

"Some books are to be tasted, others to be swallowed and some few to be chewed and digested," Francis Bacon famously said. We think this pronouncement had to do with the quality (or not) of the contents of books. But perhaps he simply anticipated the mass availability of durable fine-art editions for the very young. Such readers *are* inclined to gnaw.

* * *

PICTURE BOOKS ARE an entry point to art and illustration. They can also widen a child's aesthetic horizons. That's the explicit purpose of books for children that explore art history, which tend to start with prehistoric cave paintings and end, usually, with modern Abstract Expressionism. Claire d'Harcourt's *Art Up Close* and

Lucy Micklethwait's *A Child's Book of Art: Great Pictures, First Words* are good examples of the type. But a book needn't be about art to inculcate a taste for art. Through storybooks, children can come to appreciate different styles and traditions in an easy, natural way. For instance, anyone paging through the picture books of Chen Jiang Hong, which include *The Magic Horse of Han Gan* and *Little Eagle*, can't help but absorb the colors and brushwork of classical Chinese painting. The filigreed delicacy of Mughal and Persian miniatures comes through in books such as Diane Stanley's *Fortune* (last seen in Laura Amy Schlitz's library class), *One Grain of Rice*, by Demi, and (a huge favorite with my children) *The Seven Wise Princesses*, retold by Wafa' Tarnowska and illustrated by Nilesh Mistry. The exuberant colors and thick lines of Australian Aboriginal art fill Bronwyn Bancroft's books, among them *Kangaroo and Crocodile: My Big Book of Australian Animals*.

To me, the object here is not to instruct or demand—"You *will* look at these pictures and you will *like* them"—but in a gentle way to introduce, familiarize, and insprit. A child gazing at the fat babies, rich fabrics, and moonlit grottoes in Maurice Sendak's strange and stirring picture book *Outside Over There* may not know that he is absorbing something of the German Romantic style—but he is. The same is true when children explore Paul O. Zelinsky's paintings for *Rapunzel* and *Rumpelstiltskin*, and enter the ocher landscapes and classical interiors of the Italian Renaissance.

I talked to Zelinsky about his work. He was in college when he fell in love with the period, and he told me, "With *Rapunzel*, I was definitely showing children what I love about Italian Renaissance art. I started painting actual poses from Renaissance paintings and using them. It started with the cover, which was taken from a Rembrandt. I asked myself: Is this cheating? What is my purpose in doing this?"

The answer, like the inspiration, came from the past. "In the Renaissance, that's what painters did," Zelinsky said. "There was no expectation of everything being original. It was more the opposite. Every pose was a reference to something else. If you were educated, you knew about what was being dug up," he said, referring to the discoveries of buried Greek and Roman antiquities that sparked the classical renewal. As with fairy tales and nursery rhymes, each cultural reference could rely on a layer of previous understanding.

"Everybody is formed by what they're exposed to," Zelinsky concluded, "and if you have rich visual experiences that have meaning as a kid, how could that not feed into your ability to see and think and feel?"

*   *   *

FOR THE ILLUSTRATOR David Wiesner, creator of wordless picture books such as *Mr. Wuffles!*, *Art & Max*, and *Flotsam*, the sparks that set off great flashes came when he was a boy poring over art books at his local library in New Jersey. Wiesner found himself drawn to the subtle, intricate backgrounds of Renaissance paintings: "Look at that landscape behind *Mona Lisa*," he told an interviewer for a retrospective of his work, *David Wiesner and the Art of Wordless Storytelling*.

> I suspect that Da Vinci made that up. It looks to me like Mars or some alien place—the tiny roads, the cliffs and arches; it's fascinating.
>
> I always loved Bruegel's *Hunters in the Snow* for similar reasons. Your eye can move from the figures in the foreground, down the hill, into the town, passing all the people as they play and work until you reach the distant background. All of this is rendered with total clarity. As a kid I felt as if I was being pulled into the picture. It was full of stories.

These early encounters have a way of abiding. Growing up in an evangelical household in Florida, the writer and editor Christine

Rosen was shocked and mesmerized by a reproduction she saw in a book, of David with the head of Goliath painted on a fifteenth-century Florentine shield. "It was a story I knew very well from Bible school," she told me, "it was this bloodied head of Goliath at the feet of David, and having known the story I expected to see this victorious expression. But David's not looking victorious—he looks sort of alarmed. Like he's just felled a giant with a rock! I remember connecting to that. It was this visualization of a Bible story I knew by heart and—it just struck me. It's not, you know, the Venus de Milo, and it's not even a particularly well-known piece of art, but it just *connected*."

I had the same sort of experience myself. As a latchkey child in rural upstate New York and, later, in rural Maine, I used to study Heinrich Hoffman's dainty, grisly illustrations in *Struwwelpeter*, his 1845 collection of cautionary tales that were meant to lampoon the beetle-browed moralizers of the era, but which I took to be in earnest. I was fascinated by the incongruity of Hoffman's airy, delicate paintings and the horrors that befall the children in them. A disobedient girl plays with matches and reduces herself to a little smoking heap of ashes. A naughty thumb-sucker gets his digits lopped off by a long-legged assailant with sharp scissors. The pictures are appalling, and I couldn't look away. Like David Wiesner, I also got hold of a book of paintings by Bruegel the Elder. I scrutinized the crowded scenes of peasants roistering at wedding parties, and resting their scythes beneath shocks of wheat, and traversing the snow under bleak Low Country winter skies. It seems clear to me now, and perhaps it was clear even at the time, that studying these pictures had the effect of throwing a set of switches in my head. The time I spent with them altered how the world would look to me. Somehow, even when the figures were grotesque, they opened my way to seeing beauty.

*   *   *

"IS IT NOT the case that what makes your heart knock with fright or delight also makes your brain go tick?" This question, posed by art historian Jane Doonan, gets to the heart of the way that images can affect us. "The sensual pleasure to be derived from pictures is not something apart," she says, "but has a special role in the making of pictorial meaning."

To engage in the process of making "pictorial meaning" doesn't require a lot of time but it demands a quiet time. If we are to help children develop their aesthetic senses to the full, so that they can explore and be sensitive to the range of emotions and ideas that picture books awaken in them, we need to protect the space in which this can happen. David Wiesner, Christine Rosen, Paul Zelinsky, and I all had the luxury of making personal connections to art and imagery in an epoch before the Internet. Finding the time for that transformation is more difficult now. Life is busier, and it's harder to settle the mind. Yet the human capacity for appreciating deep and lovely things remains, and so this time is worth finding, and the effort worth making.

It takes time to look, and really to *see*, what's in front of us. With children's picture books, it can be tempting for adults to flip through the pages at the pace of the prose. Given that today's books are far less chatty and discursive than the children's fare of even half a century ago, a grown-up may zip through thirty-two or forty illustrated pages in a matter of minutes.

Looking with intensity, or "close looking," as Doonan puts it, calls for a change of gears, a downshift. "If we want to be able to make the most of a picture—to be open to it and wonder why we feel as we do in front of it—we need to look not just at what's being represented but rather at everything that presents itself, and grasp at the *how* as well as the *what*," Doonan told Jonathan Cott for his book about Maurice Sendak's "primal vision," *There's a Mystery There*. "The more you know, the more you'll be able to discover and the more meanings you'll be able to make."

As a professional, Doonan is steeped in art and its manifold in-
terpretations, but the close looking method that she advocates is
entirely accessible to the amateur, meaning, in this case, the parent who wants to help children appreciate picture-book illustrations at a deeper level.

In Doonan's primer on the subject, *Looking at Pictures in Picture
Books*, she writes, "Interpreting pictures fully involves attending
to everything which presents itself to the eye. It is not necessarily obvious that the qualities of a picture come from the artist's
style, choice of materials and compositions, nor how these pictorial means achieve their effects. Once children are shown and
told how lines and shapes and colors are able to refer to ideas and
feelings, they can explore the dimension beyond what is literally
represented. They move into partnership with the artist through
the picture itself."

What does it mean to "attend to everything which presents
itself to the eye"? It might mean considering the straightness or
crookedness of lines of ink; the depth and abundance of color, or
lack of it; the perspective or viewpoint (are we looking down on
a scene, up at it, or are we peeking at it around a corner?). What
shapes and edges predominate in the pictures? Do the people and
things depicted appear soft and gauzy, or jagged and anxious?
Each of these artistic decisions contributes to the feeling a work
gives us, and to the meaning we assign to it.

The lovely thing about engaging in this kind of interrogation
with a child is that it can be playful, experimental, and open-ended.
We may be talking here of the wise insights of art historians and
distinguished artists, but in the domestic context everything is up
for exploration. It's not a test, and there isn't a grade. It's just a
grown-up and a child, diving into pictures, noticing how they feel,
what they think, and talking it all over.

*   *   *

WHEN ENGAGING WITH our children and with creative works from the past, whether storybooks or novels or paintings or illustrations, from time to time we are going to come to depictions that startle and displease. That's understandable. The early twenty-first century is a time of intense social, historical, and aesthetic reexamination. In children's books, especially, there's a big push to widen the range of stories and storytellers so as to better represent the broad panoply of human experience. With this laudable and expansive attitude has come, however, a rising contempt for works that fail to reflect the new understanding. There's a broad temptation to view the past through the prism of our contemporary attitudes and taboos—and to find its people shameful and wanting. We believe ourselves to be so much more enlightened; but, then, people always do.

A saga of young scallywags on the Mississippi, a fictionalized memoir of the American frontier, an adventure in a turban-wearing land of a "wise, wealthy, courteous, cruel and ancient people": many classic novels for young readers reveal the existence of retrograde attitudes. The way their authors depict religion, ethnicity, skin color, and especially the sexes and their lively differences may not accord with modern sensibilities.

This is a short-term problem for novels, but it's a long-term problem for the culture. When the Wang brothers first pitched the idea of their toddler-friendly Cozy Classics, an uneasy publisher demurred: "I don't think you should do *Pride and Prejudice* because the word 'prejudice' is inappropriate for a children's book, and if we do it, we should change the title." Jack Wang laughed with incredulity when he told me the story. "We were like, do you not understand this concept? People know this book and love it. There is no way you're going to change the title of this beloved book!" That a publisher would propose changing the title of a novel by Jane Austen is a sign of the extreme skittishness of our times.

Long before our digital age, an American writer named Walter

Edmonds wrote historical fiction for young adults and won prizes for it. Accepting one of these accolades, Edmonds observed, with perspective that applies to us, "The present *is* important, but today will be yesterday in less than twelve hours, just as a little over twelve hours ago it was unpredictable tomorrow. This present that some of us get so het up about is less than the wink of an eyelid in the face of time. The past is as alive as we are."

Human cultures are deep and dynamic, as fathomless as Maria Tatar's ocean of stories, as shifting and complex as Salman Rushdie's liquid tapestry. That being the case, we are bound to encounter anachronisms. The world has changed since Sappho, since Dryden, since Hans Christian Andersen and Lucy Maud Montgomery and even E. B. White. The ways that people live and think have altered, often for the better. Still, we are wise not to assume that of all the human beings ever born, we just happen to be the ones who've got it all right.

*   *   *

OUR EPOCH IS, of course, hardly the first to be alarmed by earlier cultural assumptions. In the early nineteenth century, when reading aloud was a popular pastime of the English-speaking home, a man named Thomas Bowdler worried that the works of William Shakespeare, unexpurgated and read out loud, were inappropriate in a domestic setting. Bowdler, whose antics would give rise to the verb *to bowdlerize*, meaning to modify or remove offensive passages of text, brought forth a volume designed to forestall any unpleasantness. Shakespeare's sixteenth-century brilliance was, in Bowdler's view, marred by elements that today might be labeled "problematic." So he stripped out all the bawdy banter and vulgar double entendres, as well as Catholic references that might offend the Anglican book-buying public. The title of the resulting sanitized work is sublime in its comic prolixity: *The Family Shakspeare* [sic], *In Which Nothing is Added to the Original Text but those Words and Expressions are Omitted Which Cannot with Propriety be Read Aloud in a Family.*

It may seem silly to us now, as well as stunningly arrogant, but Bowdler had good intentions. He wanted to protect vulnerable young ears from injury. His reasoning in this respect was not so different from that of the small Chicago publisher that in 2011 printed bowdlerized editions of Mark Twain's *The Adventures of Tom Sawyer* and *The Adventures of Huckleberry Finn.*

The American Library Association, to its credit, condemned the expurgation of words that distress the modern reader. "Twain used the 'n-word' deliberately because he hated racism and he hated slavery," the director of the ALA's Office for Intellectual Freedom said at the time. "Children who read this book deserve the chance to read the book thoughtfully and in its entirety and to understand and to ask questions about why [Twain] used the word and then allow teachers, parents and librarians to answer their questions."

That's absolutely right—and we can take it a step further. Mark Twain is easy to forgive because we know his heart was in the right place. A harder case might seem to be the unenlightened or unrepentant writer, he or she who validates, through fictional characters, thoughts that have become socially unacceptable.

In Laura Miller's literary memoir *The Magician's Book*, she acknowledges the reader's dilemma: "How to acknowledge an author's darker side without losing the ability to enjoy and value the book."

Miller writes: "Prejudice is repellent, but if we were to purge our shelves of all the great books tainted by one vile idea or another, we'd have nothing left to read—at least nothing but the new and blandly virtuous. For the stone-cold truth is that Virginia Woolf *was* an awful snob, and Milton was a male chauvinist."

So too was Rudyard Kipling a colonial man of his times—and a brilliant adventure writer. Laura Ingalls Wilder's Little House books give irreplaceable insight into the lives and attitudes of white Americans during the westward expansion of the late nineteenth century. There is also no avoiding the fact that, through her characters, Wilder gives voice to prejudices that disturb us today. Modern readers flinch at Ma's open hatred and fear of Indians. Pa shows more equanimity, and even a measure of respect for the civilization his own is displacing. (Missing, of course, is perspective going the other way: In Wilder's work, we don't learn from an Indian mother her opinion of homesteaders, nor from an Indian father his views of the foreign practices of white newcomers.) There's also, in *Little Town on the Prairie*, a notorious scene in which Pa and other men of a South Dakota town appear in blackface and sing and dance in a minstrel show that all the townspeople are shown to enjoy.

Should children today know that such attitudes and events ever existed? Will they be permitted to know, as the years go by and the books are reissued? Or will problematic scenes and characters be dropped down the memory hole and made to disappear?

Ray Bradbury used his 1953 novel *Fahrenheit 451* to warn of the danger posed to literature by the ranks of the offended. In an incredible twist—considering that censorship and the destruction of troubling narratives are two of the most important ideas animating the book—Bradbury's own publisher censored certain passages behind the author's back in later editions. As the years progressed, Bradbury started to get complaints about defects in his novel, such as male chauvinism.

In an afterword to the 1979 reissue of *Fahrenheit 451*, Bradbury let fly: "It is a mad world and it will get madder, if we allow the minorities, be they dwarf or giant, orangutan or dolphin, nuclear-head or water-conversationalist, pro-computerologist or Neo-Luddite, simpleton or sage, to interfere with aesthetics."

"There is more than one way to burn a book," Bradbury wrote, "and the world is full of people running about with lit matches." It was true forty years ago, and it is true now. It's important to keep in mind that art and literature belong to the last generation and to future generations as much they do to ours. We have no more right to edit or expurgate the classics to suit our tastes than the Victorians would have had to invade the Uffizi Gallery and paint a frock over Botticelli's *Venus*.

* * *

NONE OF THIS means that, reading at home, we are obliged to speak every word out loud. In private, we can take liberties. In some cases, we may be obliged to: after the first time I read Gillian Cross's retelling of *The Iliad* to Flora, she would never again allow me to describe the death of Hector. Cross's version is far less upsetting than Homer's, yet I absolutely must stop the moment Achilles throws his spear at Hector's unprotected throat. Flora knows how the scene unfolds, but she doesn't want to be made to walk through it. When we get to that scene, I am allowed only to say, "Hector died."

That sort of responsive adaptation is part of what makes reading aloud such a vibrant experience. We can tailor the reading to the child, and to ourselves. Some parents skip past the awful moment when the hunter shoots Babar's mother. Some children, like a little boy named Theo, hide when the page comes rather than risk glimpsing the terrible scene. That's fine! There are lots of reasons to adjust as we go. Some readers skip long descriptive passages, for fear of making their audience restive. One mother I know didn't want her small daughter to hear her say, "Shut up!" even in the guise of a fictional character, so when the words appeared in the text she would instead say: "Be quiet!"

My favorite scene of literary meddling comes not from real life but from the 1960s TV show *Bewitched*, in an episode that has the redheaded witch Endora sitting down with her little granddaughter, Tabitha, to read a well-known fairy tale:

Once upon a time, a nice kindly witch lived in a gingerbread house in the forest. She wasn't bothering anyone, or causing any harm, you understand that, Tabitha?

Unfortunately, one day, two nasty little hooligans called Hansel and Gretel found the gingerbread house. Well! Without so much as a by-your-leave, the two little gluttons broke off her window ledge and ate it!

Now what do you think of that? Naturally the nice witch was concerned. I mean, who likes vandalism?

In subverting the story of *Hansel and Gretel*, the witch achieves a number of ends: while snuggling up with her granddaughter, she's sharing a classic tale and, at the same time, supplying her own idiosyncratic critique. Not to give a humorous retro TV scene too much weight, but one day Tabitha will be able to read the story for herself, and she will see the difference between the words as written and the ones her grandmother "read" to her. And she will learn something.

We do children no service in cutting them off from transcendent works of the imagination, even if it means introducing them to troublesome ideas and assumptions, and to characters we would rather they not admire. Like life itself, literature is unruly. It raises moral, cultural, and philosophical questions. Well, where better to talk about these things than at home? The human story is messy and imperfect. It is full of color and peril, creation and destruction—of cruelty and villainy, prejudice and hatred, love and comedy, sacrifice and virtue. We needn't be afraid of it. It's foolish to cover it up and pretend history never happened. It is far better to talk about what we think of these matters with our children, using books as a starting point for the conversation.

"Great art has often been made by bad people," says the writer and provocateuse Camille Paglia. "So what?"

* * *

WHATEVER CURRENTS STIR in the wider culture may be beyond our control, but we have private recourse. At home, like Chen Guang-cheng's father, we can read what we like. If we read to children, widely and without fear, we can expand their hearts and furnish their minds and, in doing so, give them a place to contend with ideas that may be too painful or awkward to discuss anywhere else.

"History, after all, is people," observed another pre-Internet writer, Elizabeth Janet Gray, author of the Newbery Medal winner *Adam of the Road*, which is set in thirteenth-century England. His-torical perspective, Gray said, "gives us a profound sense of being part of a long chain of life that went on years before us and will go on years after us, with customs and events differing in many ways but man's problems and aspirations, his grief and joys, remaining substantially the same. The realization that we are not alone, not unattended, brings color and richness to our present experiences, reinforces our fortitude, and resolves our hesitancies."

We are not stranded on a desert island with a single book to occupy the rest of our days. We can mix it up. We can read our children Laura Ingalls Wilder's *Little House in the Big Woods*, and we can read them Louise Erdrich's *The Birchbark House* and let them see—let the authors *show* them—at once how similar life is when you are a little girl with annoying siblings who lives close to the land and must engage in yucky, tiresome chores, and how different westward expansion appeared, depending on the type of house and society you occupied. The solution to problematic pas-sages in any particular book is not fewer books, but more of them.

We are fortunate to have access to more books and stories and paintings and sculptures and other art forms than people at any other time in history. We have schools and libraries and museums, brick-and-mortar bookstores and online leviathans that will sell us new volumes and old, real and virtual, pricey and at a heavy discount. Through Project Gutenberg, we can take classics off the Internet for free.

No single book has to scratch every itch. If the problem is that some literature expresses old-fashioned views, the solution is to read our children more books of every kind. The more reading, the more voices; the more voices, the more imagination; the more imagination, the more opinions; the more opinions, the more freedom of thought; and the more children engage in freedom of thought, the better.

"If all mankind minus one were of one opinion, mankind would be no more justified in silencing that one person than he, if he had the power, would be justified in silencing mankind," writes the philosopher John Stuart Mill, who, since we're cataloging faults, was also an adulterer who held grudges. "But the peculiar evil of silencing the expression of an opinion is that it is robbing the human race, posterity as well as the existing generation—those who dissent from the opinion, still more than those who hold it. If the opinion is right, they are deprived of the opportunity of exchanging error for truth; if wrong, they lose, what is almost as great a benefit, the clearer perception and livelier impression of truth produced by its collision with error."

Be not afraid. Let the stories flow. There are simple and sensible ways to convey optimism and openheartedness while acknowledging the limitations (as we regard them) of people who went before us. One of the best models I've seen for handling volatile topics with children comes from a book published almost a century ago, *A Child's History of the World*, by a Baltimore educator named Virgil Hillyer. The chapters are short, packed with nifty details, and they convey the eccentricities of the past with refreshing reason and perspective. In the book's discussion of the Hellenic Golden Age, we learn about a man named Phidias who fashioned a colossal ivory and gold statue of Athena for the Parthenon, in Athens.

"Phidias has been called the greatest sculptor who ever lived," Hillyer writes, but he did a thing which the Greeks considered a crime and would not forgive. We do not see anything so terribly wrong in what he did, but the Greeks' idea of right and wrong was different from ours. This is what he did. On the shield of the statue of Athena that he had made, Phidias carved a picture of himself and also one of his friend Pericles. It was merely a part of the decoration of the shield, and hardly anyone would have noticed it. But according to the Greek notion, it was a sacrilege to make a picture of a human being on the statue of a goddess. When the Athenians found out what Phidias had done, they threw him into prison, and there he died.

There, in one short passage, is a dose of sensible, tolerant, age-appropriate historical perspective: "We do not see anything so terribly wrong in what he did, but the Greeks' idea of right and wrong was different from ours." The past is a foreign country. They do things differently there. Children can understand this.

In this context, reading aloud can become an act of respect for the generations that came before us, of humane resistance to the roving eye of the censor. To defend classic literature is not to defend prejudice. It is to argue for sympathy, and an openness to the past as well as to the insistent present. It is to recognize that as we judge the people of former times, so shall we be judged by future ones. Attributes of our contemporary life that we think unimpeachable or unavoidable will be held up by generations yet to come as evidence of our limits, our stupidity, our profligacy, our lack of vision. The present, as Walter Edmonds said, is "less than the wink of an eyelid in the face of time."

So yes, when you are reading aloud, by all means engage with the brilliant, flawed works of the past. Why not? The books you read will enrich your children's lives long after they are grown and gone.

And then—

Well, it's over. The bedraggled baby books, the stacks of picture books shorn of their dust jackets, the dog-eared paperbacks, the hardcover classics and the new novels with their crisp pages—they all get stowed, or sold, or given away. You've fought the good fight, you've run the race to the finish. You are done.

Or are you?

Sometimes we may forget that there's more than one important dynamic in family life. The pleasure and value of reading aloud extends beyond parents reading to their children. The intellectual stimulus it brings, the emotional connection, the strange stir of shared literature; all this also happens when adults read to adults, when siblings read to siblings, and when, one day, grown-up children read out loud to their parents.

CHAPTER 8

FROM THE NURSERY
TO THE NURSING HOME
Why Reading Aloud Never Gets Old

———————

Read him slowly, dear girl, you must read Kipling
slowly. Watch carefully where the commas fall so you
can discover the natural pauses. He is a writer who used
pen and ink. He looked up from the page a lot, I believe,
stared through his window and listened to birds, as
most writers who are alone do. Some do not know the
names of birds, though he did. Your eye is too quick and
North American. Think about the speed of his pen.
What an appalling, barnacled old first paragraph it is
otherwise.

—Michael Ondaatje, *The English Patient*

Not long ago, a woman named Linda Khan was sitting by a
hospital bed in Houston, Texas, feeling ill at ease. Beside
her lay her eighty-eight-year-old father. His heart was fal-
tering. He needed surgery.

That wasn't what was bothering Khan, though. What troubled her was that all day the two of them had engaged in nothing
but depressing small talk. She loved and admired her father, and
they'd always had good conversations, but now he seemed sunk
in querulous contemplation of his predicament. He talked about
the lousy hospital food, the tests, the doctors, the diagnosis, the
potential outcomes. The scope of his once wide-ranging interests
seemed to have shrunk to the size of the room. Khan, for her part,

had a similar feeling that the world outside was becoming remote, disconnected, irrelevant.

"It is really hard to sit with a person in a hospital," Khan told me later. "They're going through so much, and it feels like there's nothing to talk about except their medical situation."

Casting around for a way to divert her father's thoughts, Khan's eye fell on a stack of books that people had brought to the hospital as gifts for him. Her father had always been a big reader, but of late he didn't have the energy or focus.

In that moment, Khan was struck with an epiphany. She picked up a copy of *Young Titan*, Michael Sheldon's biography of Winston Churchill, and started to read it out loud.

"Right away it changed the mood and atmosphere," she told me. "It got him out of a rut of thinking about illness. It wasn't mindless TV, and it wasn't tiring for his brain or eyes because I was doing the reading."

That afternoon, Khan read to her father for an hour. It was a relief and a pleasure to both of them. Reading gave the daughter a way of connecting with her father and helping him in a situation that was otherwise out of her hands. Listening allowed the father to travel on the sound of his daughter's voice, up and out of the solipsism of illness and back into the realm of mature intellectual engagement, where he felt himself again.

"He's in and out of the hospital a lot now," Khan said, "and I always read to him. It's usually military history or biography, not my usual stuff, but he has good taste. I'm happy."

For Neil Bush, the late-life hospitalizations of his famous parents, George H. W. and Barbara Bush, became opportunities to repay a debt of gratitude. "When I was a kid [my mother] would read to me and my siblings," he told a reporter in the spring of 2018. With his parents in and out of care, he said, "We've been reading books about dad's foreign policy and more recently, mom's memoir."

Bush went on, his voice thick with emotion, "And to read the story of their amazing life together has been a remarkable blessing to me, personally, as their son."

The day after he gave the interview, his mother died at the age of ninety-two.

In reading to their ailing parents, Linda Khan and Neil Bush returned to a traditional means of consoling the sick. They also joined excellent historical company. Among the many men and women over the centuries who have lifted the burden of a loved one's confinement by reading out loud, we can count the great Albert Einstein. His sister Maja had suffered a stroke in her mid-sixties and remained bedridden for the rest of her life. According to a charming account in the *New Yorker*, Maja's brilliant older brother would go up to her room in the evenings and sit for an hour or so, reading the Greeks: "Empedocles, Sophocles, Aeschylus, and Thucydides receive the tribute of the most advanced and abstract modern science every night, in the calm voice of this affectionate brother who keeps his sister company."

Einstein was a man who appreciated higher planes of thought, as we know. Perhaps it was because of his almost superhuman intelligence that he was so sensitive to the plight of an active mind trapped in an earthbound body. Years earlier, at a birthday celebration for the theoretical physicist Max Planck, Einstein had talked of the human yearning for transcendence over coarse, quotidian things:

I believe that one of the strongest motives that leads men to art and science is escape from everyday life with its painful crudity and hopeless dreariness, from the fetters of one's own ever shifting desires. A finely tempered nature longs to escape from personal life into the world of objective perception and thought; this desire may be compared with the townsman's irresistible longing to escape from his noisy, cramped surroundings into the silence of high mountains, where the eye ranges freely through the still, pure air and fondly traces out the restful contours apparently built for eternity.

A person who is limited by old age or illness may need the help of another to escape the "painful crudity and hopeless dreariness" of his circumstances. That is certainly the case with the title character of Michael Ondaatje's 1992 novel *The English Patient*. Burned over most of his body, the man is active only in his thoughts, and the young Canadian nurse who reads aloud to him keeps mangling the Kipling.

"Think about the speed of his pen," he entreats her.

The English patient's request is a good reminder that reading aloud needs to be considerate and companionable. No one wants to hear a voice droning on without regard to the words or the listener. At its best and most uplifting, the experience becomes a piece of art that the reader pulls from thin air and gives as a gift to the hearer. The artwork is composed of a writer's words and the music they make as they strike the ear, combined in the telling of a narrative that produces what radio dramatists used to call sound pictures, or "theater of the mind."

There is a performative element, too: the reader's phrasing and intonation, the pauses between words and sentences, the timbre of the voice and its warmth or chill. All these things communicate themselves in a complex aesthetic experience that is as transient as breath and as comforting, as we saw with the babies in the NICU, as physical touch.

And there is the giving of self. When we read to other people, we show them that they matter to us, that we want to expend time and attention and energy in order to bring them something good. In earlier chapters we saw the way that this works in young families, how shared stories can bind parents and children together and sweep them away in a lovely neurochemical tsunami. The same magic is at work when everyone involved has long since grown up.

* * *

IT CAME AS an unpleasant surprise to Lauri Hornik when her daughter announced, at the age of ten, that her mother's services as a reader were no longer required. Ruby wanted to read on her own, at the faster speed of her own eyes. "That was a very sad moment for me, but I had to allow it," Hornik told me.

Three years went by, and reading aloud seemed a pleasure that had been relegated to yesteryear when a chance confluence of things occurred: Hornik and her partner, Peter, were embarking on a five-hour drive through the Adirondacks; cell coverage was poor; and Hornik had just got her hands on the unpublished manuscript of John Green's hotly awaited 2017 young adult book *Turtles All the Way Down*. As the president and publisher of Dial Books for Young Readers, Hornik was one of the first people in her industry to see Green's latest novel since *The Fault in Our Stars*. She was longing to dig in, but felt it would be boorish to read to herself while Peter did all the driving.

"I couldn't wait," she said, "so I started reading it out loud to Peter, and I think we went for several hours of just me reading it to him, because we couldn't stop. Reading it aloud was such an experience of being *in* the head of the main character. If you're reading a first-person account aloud, then all the more you *are* that character. So it was very deep and meaningful to me. I think also, when you're not reading aloud, you're often sort of skimming, so that each sentence doesn't necessarily resonate in the way that it does when you're reading it out loud."

That resonance is best achieved when the reader takes his time. As the English patient warned, it is all too easy for eyes to be "too quick and North American," zipping along, skittering and jumping and taking in meaning by the glance. The ear demands a steadier pacing. Indeed, by its deliberate nature, reading aloud forces us to interact with writing in the way that Vladimir Nabokov said we ought.

"Literature must be taken and broken to bits, pulled apart, squashed," the novelist once wrote; "Then its lovely reek will be smelt in the hollow of the palm, it will be munched and rolled upon the tongue with relish. Then, and only then, its rare flavor will be appreciated at its true worth and the broken and crushed parts will again come together in your mind and disclose the beauty of a unity to which you have contributed something of your own blood."

When we experience a text out loud, word by word, we give weight and value to language even as we are subsumed by it. The effect is not just pleasurable. The companionship and intellectual stimulus reading aloud provides seems also—in real ways—to promote the health and well-being of both the reader and the listener.

* * *

ONE JUNE AFTERNOON in North London, not so long ago, half a dozen elderly people sat in comfortable chairs at two round tables on the fourth floor of a facility for the frail and aged. Outside, clouds hung flat and heavy in the sky. Inside, the meeting place was pleasant and hotel-like, with soft carpeting and stand-alone bookcases. There was no medicinal smell, no sign of the violent catastrophe that had shaped these people's lives.

A younger woman, Kate Fulton, had just served hot cups of tea all around, and now she handed out sheaves of stapled photocopies.

"Right," Fulton said as she slid into her chair, "we've got a story by Doris Lessing,"

"Doris Lessing," someone echoed.

"I think she went to the same school I did," came a languid voice.

Fulton smiled. "Just to remind, there are no rules other than—"

"Listening."

"No, not that one. No reading ahead!"

Amusement ran around the tables. Everyone knew this. It was being explained for my benefit.

"The whole point of this group is to be in the moment with the literature," Fulton said to me, "so that when we stop, we see where we are. And we assess. We're only at a moment in time."

She turned back to the group. "Everybody ready? Right, I'm going to start. 'Flight,' by Doris Lessing. 'Flight.'"

"Is she still alive?"

"No, she has died," said Fulton, "so it's an homage to her today."

After a pause, she began to read. Her words were loud and crisp. She watched carefully where the commas fell.

Above the old man's head was the dovecote, a tall wire-netted shelf on stilts, full of strutting, preening birds. The sunlight broke on their gray breasts into small rainbows. His ears were lulled by their crooning, his hands stretched up towards his favorite, a homing pigeon, a young plump-bodied bird which stood still when it saw him and cocked a shrewd bright eye.

"Pretty, pretty, pretty," he said.

Most of the old people at the tables sat unmoving, their faces angled down as they followed the story in their booklets. Behind dark glasses, one woman, who was blind, had her face turned toward the reader. The room was filled with a kind of quiet, concentrated intelligence. There was an open and interested knowingness in this group of people whose accents—Gallic, Teutonic, English expatriate—hinted at the postwar diaspora that had brought them all to the predominantly Jewish neighborhood of Golders Green. All were Holocaust survivors, with backgrounds as varied as their accents. Several of the women had PhDs, and one had taught literature at a university. The sole man, a cheerful South Londoner, described himself as "Not clever, me. Bottom of the class!"

Whatever their separate lives had been, they now spent an hour and a half every week sitting together and enjoying literature out loud.

"'Her hair fell down her back in a wave of sunlight . . .'"

As Fulton reached the end of the first page, there was a soft, prolonged flapping as people turned to the next sheet.

"'. . . and her long bare legs repeated the angles of the frangipani stems, bare, shining-brown stems among patterns of pale blossoms,'" the reader continued, describing the teenage granddaughter who upsets her dovecote-tending grandfather in Lessing's story. The man sees the girl, and puts the pigeon away roughly. Wary, the two characters greet each other. There is some cause of strain between them.

At this point, Fulton looked up from the text.

"Okay," she said, "where do you think we are?"

"In the country?"

"By a railway line."

"What do you make of the old man, so far?"

"He's a bit moody."

"Moody? What makes you think that?" asked Fulton.

"Oh, well, he suddenly decides to shut his favorite bird in a cage."

"Right. He was fine before, but there was a sudden change in his mood, wasn't there? When did that happen?"

"When he saw his daughter—no, his granddaughter—swinging on the gate," said the man.

"Yes, what do you think he's feeling?" Fulton asked. "Imagine him, he's sitting—put him in his space—he's in the countryside, I don't know where it has red soil, anybody know? Highlands, lowlands, somewhere where there's red soil?"

*   *   *

WHEN I VISITED Golders Green, Kate Fulton had been reading aloud with the group every week for five years, having given up her career in law to "nourish the soul rather than the bank balance," as she put it. Her group ran under the auspices of The Reader, a national charity founded in 2002 by Jane Davis, a professor at the University of Liverpool who wanted to bring great literature out of the ivory tower and into places where ordinary people live, and in particular to places where they suffer.

The Reader sponsors hundreds of groups in Great Britain, by no means primarily for older people. There are groups for schoolchildren and foster children, teenagers and prisoners, patients in neurological rehabilitation units and inmates of psychiatric wards, recovering drug addicts and people with Alzheimer's disease, as well as for nurses and other stressed-out caregivers.

Whatever the makeup of the group, the protocol is the same. The facilitators are trained to read in calm, modulated tones rather than with a lot of theatrical acting-out, so that the author's language can come through in a clean and unadorned manner. Occasionally other people at the table will take a turn reading that week's material, which in most cases consists of a short story and a couple of poems.

"I have to make sure the imaginative places that I take them from are varied," Fulton told me.

We went to a magic shop with H. G. Wells the other week. We're walking down some Russian street with Chekhov, we're somewhere with Maupassant, or actually we're just having a door fixed with Rose Tremaine.

They enjoy it very much. I had one lady who never spoke. Remember, these people have had quite a difficult past, even if they were only in camps until they were eight. It can affect you hugely. This lady had been evacuated, but I didn't know because she never spoke. We read a poem, Longfellow's "The Arrow and the Song," and it's all about friendship. She suddenly said: "That's a song, you know, Kate."

She'd never spoken before! I said, "Would you like to sing it to us?" And she did! She started crying. The tears were pouring down her face, and she said, I was evacuated in the war and we were at a fire station and I haven't heard that song since then and you've just given me back my childhood.

When you're in these kind of groups, talking about stories, anything can come up. You just don't know what's coming.

"It makes me *think*," the blind member of Fulton's group said. "You use your brain and you find things that interest you that you normally don't talk about. You get an insight into different stories. It's surprising what you can learn about other people, too. It's not as though you're in school and being taught. It's a friendly relationship."

"It's interaction with other people," the man put in.

"It brings literature to life," said another woman. "You hear something and you discuss it. You can put yourself in the protagonist's shoes. It's stimulating. Otherwise, you look at the four walls or watch television or something like that."

*   *   *

THERE'S REASON TO believe that reading groups of this sort offer the people who take part more than just vague, anecdotal benefits.

"On the emotional level, it's just wonderfully nourishing," said Paul Higgins, an early volunteer for The Reader who eventually became the organization's first salaried employee. "Not all of us have had that wonderful, almost primordial experience of being read to as children, and soothed. Interestingly, given the opportunity, people, particularly older people, often say, 'It's so relaxing.' Loving-kindness is what people experience, through the literature, through the network that builds up, the lifeline that builds up week to week. Kindness, love, and beauty. That's what hits people. That's what's so amazing about it."

In a 2010 survey in the UK, elderly adults who joined once-a-week reading groups reported having better concentration, less agitation, and an improved ability to socialize. The survey authors attributed these improvements in large part to the "rich, varied, non-prescriptive diet of serious literature" that group members consume, with fiction encouraging feelings of relaxation and calm, poetry fostering focused concentration, and narratives of all sort giving rise to thoughts, feelings, and memories.

There may be another desirable long-term benefit of enjoying serious literature. Recent research at Yale University finds that people who read for pleasure live an average of two years longer than nonreaders, and that, further, reading books seems to have a greater protective effect than reading newspapers or magazines. "This effect is likely because books engage the reader's mind more," explained Yale's Avni Bavishi. "Cognitive engagement may explain why vocabulary, reasoning, concentration, and critical thinking skills are improved by exposure to books." Literature, she said, "can promote empathy, social perception, and emotional intelligence, which are cognitive processes that can lead to greater survival."

More exciting still, perhaps, are the effects of reading aloud on people who have Alzheimer's disease. A 2017 paper from clinicians at the University of Liverpool hints at huge potential promise for not only the 800,000 men and women with dementia in the UK but Alzheimer's sufferers everywhere.

"Reading a literary text together not only harnesses the power of reading as a cognitive process: it acts as a powerful socially coalescing presence, allowing readers a sense of subjective and shared experience at the same time," the study's authors wrote.

That is easy enough to appreciate, of course, but there's more: "Research suggests that the inner neural processing of language when a mind reads a complex line of poetry has the potential to galvanise existing brain pathways and to influence emotion networks and memory function."

(As Cornell's Morten Christiansen told me, "Experience and use of language do matter throughout your life. Language is a bit like a muscle. It will atrophy if you're not using it." In Japan, with its large population of elderly people, clinicians are exploring how a bit of daily reading and math can sharpen cognitive skills that have been dulled by age and lack of use. In 2016, researchers at Tohoku University enlisted a group of healthy elderly people to undertake "learning therapy" for six months. The therapy involved the volunteers performing simple math calculations and reading short passages of Japanese prose out loud. By the end of the experiment, many of the subjects had experienced cognitive improvements.)

For many people who get together every week to read aloud, part of the pleasure comes from the sheer physiological relief of being in the company of others. Men and women who live by themselves, or who are confined to hospitals or nursing homes or prisons, may have only rare contact with others, or contact that is principally practical, transactional, and hierarchical. They may not have many opportunities to engage with other people as equals, let alone to experience poetry or short stories out loud. With affecting candor, an inmate of one of Britain's most notorious prisons described how participating in the Reader sessions felt to him: "It is almost as if literature 'raises the bar' and leaves the reader feeling like an explorer of a different world, or at least privileged to have glimpsed into another realm that is 'otherworldly.' The group's overall relaxed and soothing atmosphere seems to draw me near and *fill some sort of need in me I did not know was there.*" (Italics mine.)

*   *   *

MODERN LIFE CAN be a lonely, isolated affair. With the advent of the digital revolution, loneliness as a cultural phenomenon appears to be intensifying. By one recent assessment, rates of loneliness have doubled since the 1980s. Today in the United States, upward of 40 percent of adults suffer from some degree of isolation. We are social animals, as Aristotle said. Feeling disconnected and alone can take a grievous toll. People who feel lonely are three times as likely to report symptoms of anxiety and depression, according to a recent Danish study. Lonely hearts face double the mortality risk from diseases *of* the heart.

"The world is suffering from an epidemic of loneliness. If we cannot rebuild strong, authentic social connections, we will continue to splinter apart," former US surgeon general Vivek Murthy wrote recently in the *Harvard Business Review*. This epidemic is doing damage not only to minds and hearts, he argued, but to our bodies: "Loneliness causes stress, and long-term or chronic stress leads to more frequent elevations of a key stress hormone, cortisol. It is also linked to higher levels of inflammation in the body. This in turn damages blood vessels and other tissues, increasing the risk of heart disease, diabetes, joint disease, depression, obesity, and premature death. Chronic stress can also hijack your brain's prefrontal cortex, which governs decision making, planning, emotional regulation, analysis, and abstract thinking."

It is an awful catalog of suffering, given how simple and inexpensive the means of relief. Reading has an astonishing power to salve and console, lifting the lonesome from their isolation and offering reprieve to the sick from the exhausting weight of illness.

We are not even the only species to benefit. Dogs do, too, which is why, since 2014, volunteers at the American Society for the Prevention of Cruelty to Animals have read to the animals to help them recover from trauma.

"Ten or fifteen years ago, I was essentially the only person who worked with the neglect and abuse cases," said Victoria Wells, the organization's senior manager for behavior and training, when we met at ASPCA headquarters in Manhattan.

There were times when the dogs were isolated in quarantine because they came in and they were ill. They couldn't get up, because they were being treated for severe injuries. I wanted to interact with them somehow, but I couldn't take them out and work with them in a physical way. It wasn't good for them medically.

So I used to sit with them, in front of their kennels, and play guitar and sing. I used to play the Beatles. I noticed that the dogs who were very fearful, in the back of their kennels shivering and cowering, would slowly creep forward to the front. They would appear to be listening and they would become very relaxed.

The dogs' response to music led in a natural way to the idea of reading aloud. It was a practical means of allowing a larger number of volunteers to minister to recovering animals without having to interact with the dogs in a direct way that might be intimidating. Wells and her colleagues worked out a careful protocol to minimize stress. Today's volunteers are trained to use an even, reassuring tone of voice, and to sit so that they aren't directly facing the dogs, lest they seem confrontational. The choice of material is up to the readers. Some volunteers keep the animals appraised of current events by reading the newspaper, some choose children's books, and others prefer adult fiction. On the day I stopped by, a retired opera singer was reading the 1967 sci-fi thriller *Logan's Run* to half a dozen dogs housed in a row of clear glass enclosures. I noticed a lot of noisy barking when she started, but soon her voice settled on them like an audio blanket, and the animals subsided into restful poses.

"The dogs really enjoy the reading," Wells told me. "I think the fact that it's nonthreatening but it's attention, all the same, is what is most beneficial to these dogs. We noticed that it really does assist in the standard behavior treatment. The dogs are much more receptive to us, they seem more comfortable in their kennels in general (because it's sort of like being in a fishbowl, and when people loom over them it can be very intimidating), but it really prepares them for people having to walk by, and adopters coming and looking at them and potentially taking them home. I think it's that soothing, even tone of voice and the presence of somebody to keep them company that really, really benefits them."

If even dogs flourish when we read aloud to them, it's hardly surprising that people do, too, whether it happens by design, as with the British reading groups, or through a moment of serendipity in a hospital room or on a long drive through the Adirondacks.

For adults, literature shared by the voice becomes an opportunity for encounter, companionship, and self-discovery. It's a balm for the lonely heart and a means of escape from surroundings and confinements that may be as much mental as physical. It offers connectedness both in the moment and, in a deep way, with the full richness of human experience. "You read something which you thought only happened to you, and you discover that it happened one hundred years ago to Dostoyevsky," James Baldwin once reflected. "This is a very great liberation for the suffering, struggling person, who always thinks that he is alone. This is why art is important. Art would not be important if life were not important, and life is important."

*   *   *

LITERARY ART HELPS us live longer, and enjoying it together out loud makes us smarter, happier, and more contented. It may even be—we can perhaps extrapolate from the Tohoku University research—that being the reader, the rhapsode, is in itself good for the body and soul.

The second-century Roman doctor Antyllus thought it was. Antyllus recommended daily recitation to his patients as a kind of health-giving tonic. He had the fanciful idea that some arrangements of words were more wholesome than others: "Ideally one should declaim epic verse from memory, but if this is not possible it should be iambic verse, or elegiac or lyric poetry. Epic verse is, however, the best for one's health."

I somehow don't see hordes of wellness enthusiasts rushing to take up the recitation of epic verse as a way of fending off aches and pains, but those of us who adore reading aloud will be the first to say that it *does* feel as if something salubrious happens when we hold forth. The human voice is potent, and when it is put in service of beautiful writing, the effect can be a delight from the inside as well as the outside.

Reading to a spouse or a sibling or a parent might seem like a bit too much effort—so far outside the normal range of most people's regular activities as to be eccentric and a little peculiar. Linda Khan told me that right before she started to read the Churchill biography to her father in the hospital, she was tempted to put the book down again. It felt odd and even improper to presume to read to a man who, for her entire life, had always been strong and independent. She did not want him to feel patronized. Her fear was misplaced, as we know; they both ended up loving the experience. Like so many others who brave the momentary weirdness of reading to another adult, they were, to borrow a phrase from Wordsworth, surprised by the joy of it.

Who wouldn't want that? One night years ago a friend of mine wandered into his family's living room after supper and picked up a paperback copy of Michael Shaara's Civil War novel *The Killer Angels*. Without thinking much about it, he started to read the preface out loud. Immediately he was joined by his eldest son, who was about twelve at the time. A moment later his wife came in, followed by the couple's two young daughters, who at six and eight were not perhaps the target audience for an introduction to Robert E. Lee and Joshua Chamberlain but wanted to be part of a family moment. Within a few minutes, everyone seemed so comfy and engaged that my friend kept reading. It went on for an hour that night. He picked the book up again after dinner the next night, and the next, until he had finished the book. That experience of enjoying literature together, my friend told me, remains one of his family's happiest memories, one they still talk about when they all see each other.

"I wish I could tell you that it became a tradition to read together every night after *The Killer Angels*," he told me, "But we never did it again. I don't know why. I guess we just got busy."

My friend is wistful about it now. He and his family were surprised by joy—drawn together for enchanted hours with a book that had them all in thrall—and then they let it slip away.

CHAPTER 9

THERE IS NO PRESENT
LIKE THE TIME

———•———

Read to me!
Any time! Any place!

—slogan of the Family Reading Partnership, Ithaca, NY

O nce upon a time there was a modest house in the suburbs of a small American city, and in that house lived a middle-class family, the Rashids. The parents were Julie and Alex, and at the time of our story they had three children: ten-year-old Eva, six-year-old Joseph, and twenty-one-month-old Ethan, a compact bundle of muscle and energy known as "the baby."

Julie and Alex had heard that it was a good idea to read to their children, and they planned to do it . . . someday. But with the tumult of life in a busy extended family—both parents belong to large local clans, hers half Greek and his 100 percent Syrian—and with work and school and with iPhones and iPads and the big-screen TV in the living room, not to mention the inevitable upheaval of the baby's arrival, neither parent ever got around to it.

Being willing, but so far unable, the Rashids were perfect test subjects for a reading experiment. They agreed to undergo a three-month trial that would take them, overnight, from reading nothing out loud to reading every day. They promised me that they would turn off their tech and read for at least half an hour each time, with the understanding that we all fall short of the mark and that they might miss a session once in a while. What would happen? Would Julie and Alex find it a chore? Would the children squirm and bolt for the TV? Could the baby pay attention for five minutes, let alone thirty? Would the parents see any change in their children's vocabulary? Like a conservationist introducing wolves into the wild—except, I suppose, it was the opposite—I wondered what effect the alteration would have on the inhabitants of this plugged-in ecosystem, and hoped for the best.

"We're really excited," Julie said, when I showed up on their doorstep one Memorial Day weekend with two bags of board books, picture books, and chapter books to help get them started.

Inside, the atmosphere was happy and electric. Joseph dove for the bags and began grabbing books willy-nilly. "What book is this?" he said several times, not waiting for an answer and not really even looking as he tossed one volume after another onto the shag carpet of the family room.

The baby was nonstop action, treading on the books his brother had thrown, pulling off dust jackets, rolling around holding his feet, angling for my digital recorder wherever I put it, and at one point working his way behind me so as to thump me in a friendly way between the shoulder blades. Quiet and smiling, Eva leafed through some of the offerings, shushed Joseph, who tended to interrupt, and rescued picture books from the baby's investigating jaws.

"It's something I've always wanted to do and never made the time for," Julie explained. "So now we're going to implement it. It's on the calendar. We have a set schedule. It's like we're signed up for the next activity, and we're all on board. We're excited about it."

"It's going to keep us accountable," Alex agreed.

While the grown-ups were planning, the boys were in riot mode. Joseph had pulled the cover off *A Butterfly Is Patient*, spun the denuded book onto the sofa, and turned to find a fresh victim. Without missing a beat, his mother restored the cover to the book and was saying to her husband, "I'm thinking of implementing the book reading at an earlier time, because you're home by the seven or seven thirty time slot."

"Just have fun with it," I said, "No need to be strict about who reads, or when, or where."

I had meant to tell the Rashids about dialogic and interactive reading, but in the moment decided against it. They were starting from scratch and taking it all so seriously that I was afraid of overwhelming them. I did suggest that they try reading while the boys were in the bath, and urged Julie and Alex to let the children move around and play with toys during the reading if they wanted to. It was hard to imagine this frenetic troupe sitting still.

"Wonderful," said Julie, raising her voice over Ethan's bawling. "We're going to start June first, and we're going to give you three full months."

*   *   *

THREE MONTHS LATER. . . .

No, wait.

Let's leave the Rashid family to get on with their experiment for the moment. In the meantime, we can talk about how to create an enchanted hour, or even just a magical twenty minutes, at home.

Where do we start?

Start small. Start where you are. Start today! There is no need be heroic and commit to an endless future of reading aloud, or to a three-month trial period, or even to a full sixty minutes. Just begin. Pick up a book, or a magazine, or a cereal box, and try reading it out loud to someone you love.

No, really, what should I do first?

Okay, fine. First, power down your technology. Silence your phones and, if you can, put them far enough away that you won't see them or hear if they buzz. Give everyone the psychic space to engage with the words and story, and with each other. Devices mess with our ability to be present in the moment, as we know; they also disrupt the development of joint attention and emotional connectedness.

A 2017 study, "Learning on Hold," gives an unsettling glimpse of the degree to which cell phones, in particular, tend to sidetrack parent-child interactions. Moms participating in the study were given the task of teaching their two-year-old children two words, one at a time. The women all had their cell phones with them, and the clinicians made a point of interrupting half of the teaching sessions with a phone call.

"It rings. The mother picks it up. And she's been engaged with the kid, and it's like she's breaking the set with the kid," said the University of Delaware's Roberta Michnick Golinkoff, a coinvestigator on the study. "And you know how you are when you answer the phone, your face goes flat? So her face goes flat, she looks away from the kid, and then the kid doesn't want to come back. Even though she gets off the phone and she repeats and repeats, and tries and tries, and the kid won't learn the word." When the mothers and children were interrupted, the children did not learn the words. When the pairs were left to themselves, without the interrupting call, the toddlers learned the new words.

For reading aloud to work its full magic, it's best not to have competition from technology. So, yes, please, turn it off.

Are you saying I can't read aloud to my children from my tablet?

You can, but it's problematic. As with cell phones, tablets represent whole worlds of potential interruption, investigation, and short-term stimulus. *You* know that diversion is just a finger swipe away, and so does your child. Distraction is simmering there, just under the surface of the e-book, and it can mitigate against full immersion in the reading experience.

"Even with technology in our lives," as Dr. Perri Klass points out, "there *has* to be some time when we put it aside. And the more stressed the family is, or the harder your child is, which is a very real issue, the greater the temptation to take refuge in the screens."

Reading from a tablet presents no difficulty in a setting where one adult is reading to another, but when there are children involved—assuming that you want to maximize the social, emotional, and language benefits—it's best to skip the screen. Just go with a book.

What if he hasn't even been born?

Wonderful! Then you have the chance to get comfortable with reading aloud and get yourself into the habit before he arrives. The sound of your voice may have a salutary effect on his brain before birth—though as Georgetown Hospital's Dr. Abubakar says, we don't know exactly what influence it may have. Once he has arrived, though, we know that his brain networks will activate when you start talking.

What if my "baby" is already in the fourth grade? Or sixth? Is it too late?

Never. While it's probably too late to read baby books together, it is not too late to take baby steps toward a goal. Find ten minutes on either side of dinner, say, or at bedtime, or at any point in the day when you and your child will naturally be in each other's company. Again, start small. You could try reading a poem, or a news story, or maybe something your child has been assigned (or has already read) at school. Do the same thing at roughly the same time for ten minutes the next day, and build from there. If you never progress beyond ten minutes, so be it. But you'll still have built a lovely point of engagement into each day. Even worldly sixth-graders like to have warm, positive attention from their parents.

"There are few things that feel to a person like they are more cherished or taken care of," says LeVar Burton, the longtime host of the PBS show *Reading Rainbow*. "I mean, there's being fed, and then there's someone reading to you."

That's beautiful. But I am not LeVar Burton. I don't have the voice of a trained actor. How am I going to get my kids to pay attention?

Try to enjoy yourself. The more you enter into the reading, the more persuasive your example. It helps to have a great book, of course (you'll find lots of them listed at the end), but if I've convinced you of anything, I hope it's that reading aloud is a complex and bountiful experience. The story you select is only one of the ingredients. You are another, and then there's your child, with his mind lit up and his senses engaged in listening, and perhaps looking at the illustrations. (Or he may be across the room playing with Lego or drawing with markers while you read—that's fine, too.)

Entering into storytelling mode has a mesmeric power of its own. The sight of a parent or teacher sitting down with a book attracts young children like iron filings to a magnet. I once saw the same thing happen with a much older crowd in the improbable, fever-dream setting of a Florida theme park. Late-afternoon throngs were moving through the faux streets of Diagon Alley, which is part of the Harry Potter–themed portions of the Universal Studios enclosure, when a woman stepped out onto a low landing and called out, "Gather 'round!" The storyteller was wearing drab robes, so there was a small element of pageantry, but what was really striking was the speed with which people responded to her appeal. In a nanosecond, she had drawn a large and attentive audience. For the next twenty minutes, no one moved while the woman and several other performers recounted J. K. Rowling's story "The Three Brothers." When the final applause died away, it was clear from people's faces that they were going away refreshed.

You might think a theme park would be the last place on earth that the unadorned voice could twang a chord in the human heart. Yet the crowd's heartfelt response was a good reminder of the power of the spoken word and people's natural love for stories. Both work in your favor, when you read at home. You do not have to be LeVar Burton. There is no absolute need for theatrics. All you have to do is let the words unspool. Sentence by sentence, they will cast the spell.

Having said that, it is also true that the more invested you are as a reader, the more persuasive the experience will be for everyone. Silly voices and comic accents can add a lot to the pleasure. If you're reading nursery rhymes, or comic verses such as Edward Lear's *The Owl and the Pussycat* or the Oompa-Loompa choruses from *Charlie and the Chocolate Factory*, you might like to sing the words rather than read them.

No. I don't want to sing.

No worries! It's up to you. Small children do love singing, though, and when the reader is playful with the text, and with them, it opens the way for them to be playful, too.

"My daughter once caught her twin two-year-olds creeping away from their daddy's lap as he struggled to read to them," one grandmother confided to me.

> And, so you know, their daddy is a partner in an immensely de-manding law firm. He's hugely literate and the son of professors.
>
> She said, "Sweetheart, you are reading *The Gruffalo* as if it were a legal brief! Get into it. Don't recite! Entertain them!"
>
> It hadn't occurred to him.

Reading can take on an element of adventure by changing the location. Earlier I mentioned a father's technique of reading seafaring books while he and his kids were crammed in "belowdecks" on the bottom tier of a bunk bed. One mother dropped everything to take a picnic, an almost-due library book, and her daughters to a grassy hillside near their house. In the fresh air and sunshine, they read out loud in turns until the book was done. "This memory made my family realize how much it is the little things, like taking time to read together on a sunny hillside, that will be remembered and treasured for years to come," one of the daughters told me years later.

There must be *something* about reading aloud that is not, I don't know, perfect and idyllic . . .

There is. It can be incredibly difficult to make the time. There's no getting around that.

"I think it is *great*. I love every minute I am reading to them, I really do. I love the coziness of it, the stories," a woman named Carolyn Siciliano told me, after she and her husband introduced reading aloud into their cheerfully chaotic household. "But I do sometimes feel guilty. Everything takes a long time, and they are kids that need sleep. So there are times when I feel frustrated and disappointed. They'll be like, we don't get to read? I'll be like, ugh, but we can only read for twenty minutes. For us, it's really tough to fit it in."

Some parents have a hard time staying awake. "I never do *not* fall asleep," one mother said. Another confessed: "I loved the Beverly Cleary books, and I was so excited to read them to my kids, and—I don't know, but every time it's like taking a sleeping pill."

Yet another mother sent me a photograph of her youngest daughter, who had been waiting so long one night for someone to come read to her that she'd crashed out underneath a stack of picture books. There the child lay in bed, her little mouth open. Zzzz.

My son, Paris, told me that he was glad we'd read together for as long as we had, but that when he was a grouchy preteen, story time had sometimes felt like an obligation. "When I got older and thought back about it, it was a good memory that stuck with me, it was one thing I could always count on, reading in the evenings," he said. "It was never a bad memory, except—"

Well, there was that one time that I was reading *Pinocchio* to him and the girls from a beautiful edition illustrated by Roberto Innocenti. One of the final pictures comes as a shock. When you turn the page, you're confronted by a terrifying red, toothy maw.

"The girls were scared of the picture of the shark at the end," Paris went on, "And I was like, 'Wah!' and I pushed the book at them, and you sent me to my room.

"Then I drew a cool pirate's map on my door, and you were mad because I did it with a Sharpie."

I guess we'll try to read aloud, then.

"Do. Or do not. There is no try," as Yoda says. If you want you and your family to accrue the riches that reading aloud offers, I'm afraid you *will* have to make it happen. Does that sound daunting? Are you thinking: I have enough to manage and balance without adding another responsibility?

I promise: it can become a habit faster than you would believe. In Adam Alter's book about tech addiction, *Irresistible*, he describes a subtle technique that helps people change their routines. It involves adjusting the language of self-determination. "I'm trying to stay off social media" may be a true statement, but the "trying to" leaves a lot of unspoken wiggle room. According to Alter, it is more effective to close off your own avenues of escape by saying "I don't use social media" or "I don't go on social media." I see no reason why this strategy would not work for reading aloud. Instead of saying, "Well, we're trying to get around to start reading aloud," try "We read aloud every day," or "I never miss an evening of reading with my kids."

Is it best to read in the evenings?

You should read whenever it feels right. If you're at home all day when your children are small, there may be lots of odd, fugitive moments when you could pick up a book. If you're working outside the home, it may be easier in the early years to piggyback reading onto other hands-on activities, such as during breakfast or at bath time. ("I used to read to my son when he was in the high chair," one mom told me, miming the act of holding a book in one hand and spooning food with the other.)

As children get older, bedtime is usually best for corralling purposes. That's what worked for my family. After the chaos of dinner and the maelstrom of getting children bathed, brushed, and into pajamas, to reach the story hour did feel, as I said at the beginning, like crawling onto a life raft. The seas were rocking, everyone was still a bit damp and windblown, but we were safe and there was a heavenly feeling of rescue and recovery.

But my kids are years apart in age. Are they supposed to listen to me read the same books at the same time?

Why not? Maybe. It depends. You may need to experiment to find what suits your particular basket of personalities. Whether you read to everyone at once, or to each in turn, also may evolve with time. Over the years, I've catered to all sorts of age configurations. There were jolly times when everyone piled in together and we'd read a stack of picture books and then a chapter or two of a longer story for an hour before bed. Everyone got what they needed: the familiarity, reassurance, and visual pleasure of picture books (which older children like, too), along with the vocabulary- and imagination-stretching of chapter books that required the children to create scenes in their own heads.

We had one intense span of time, lasting maybe two years, when I'd spend forty-five minutes reading picture books to my younger children, and then go downstairs to read *Kidnapped*, *Treasure Island*, *The Swiss Family Robinson*, and *20,000 Leagues Under the Sea* to the older two for another forty-five minutes.

I guess you didn't have much of a social life.

No, it was fine! If I had to work after dinner, or if my husband and I had plans, I'd read while the children ate an early supper. But it does take sacrifice, there's no getting around it. Considering the good that comes of reading aloud, I'd say it's cheap at the price. Naturally there were times when we only read a couple of picture books, or I punted until breakfast, or it was so late that I turned off the lights, saying, "The movie was your story tonight."

I may be a zealot, but I'm no saint.

You keep saying "I." Didn't your husband ever read?

He did, but rarely. He worked long hours when our children were young, and seldom got home before they were in bed. Also, because I like to read aloud, I enjoyed the job. He would join us when he could, though, to listen.

Does it matter who reads aloud? Does it make a difference for kids whether a man or a woman is reading to them?

No and yes. The great thing is for children to have reading time, and the more voices, the merrier. Usually the person who likes reading aloud the most—the thwarted thespian, for instance (ahem), who has no other outlet for goofy voices—will take the lead. As a fellow enthusiast said, "Reading aloud to people, if they'll let you do it, is just about the most fun you can have."

Academic research suggests that men and women tend to bring different qualities to read-aloud time, though I'm told that the variation between people as individuals is probably far greater and more meaningful than those between the sexes. Researchers at NYU have observed that fathers tend to talk more with children about math and numbers. When sharing picture books, for instance, they're more inclined to ask a child to count the blocks or teddy bears. Mothers and fathers both give children lots of cognitive stimulation when they're reading, but their emphases may differ.

I heard a hilarious example of this. "We had a book called *The Maggie B*, by Irene Haas, about a girl who takes her little brother out on a cozy little boat," a mother of six told me.

She tidies the boat and makes it shipshape. She sings lovely sea chanteys to little James. She makes him a warm, delicious dinner and battens down the hatches when the weather gets bad. I'd read it together with the kids, all snug on someone's bed, and we'd sing the chanteys, and we'd talk about the dinner, and we'd slowly turn the pages. I loved the story. The kids loved the story.

My husband did not love the story. He thought it was nothing like what a sea story should be. So whenever the kids asked for *The Maggie B* from him, he transformed it into *The Curse of the Maggie B*. His version came complete with pirates, and howling gales, and burbling pots of sea stew, and a narrow escape from certain death. Of course, the kids like his version much better.

Another father was so tired of reading the same picture book to his three children—a book they adored—that he rewrote the text. The whole family wound up memorizing his deeply subversive version. "Thomas the Tank engine wouldn't stop horsing around," recited this man's wife. "He was always off the rails in junior high, ended up in and out of rehab, landed in jail for a brief stint, couldn't hold on to a job, got into more than a little trouble in Vegas . . ."

There's another intriguing variation in reading styles that pertains to the smallest listeners. In research settings, clinicians have noticed that the adult who spends the most time with a young toddler tends to be more in tune with the child's emerging language skills. When the child answers a question with indistinct or imperfect phrasing, that grown-up easily understands: "Yes, honey, that's the elephant!" An adult who spends more time away may need clarification, and may have to ask, "What? What did you say?" This encourages children to try harder to speak clearly and make themselves understood. So having different family members read aloud will stretch both a child's comprehension and his ability to express himself.

Reading can also be the one time in a day that the busier parent has a chance to bond with the child, or children, in a quiet time set aside for themselves. What may start as an obligation can become the most treasured hour in the day.

Do I need to show my young child the words as we go along?

Why not? Looking at words can be part of the pleasure, and it can be educational, too That's especially true with stories that rhyme. As your child is about to shout out the last word of a line of rhyme ("the cat in the—?"), by all means point to the word ("hat!"). But there's no need to make an ordeal of it.

"When my kids were little," said Laura, a mother of three, "we tended to read the same picture books over and over. So it was fun when they could 'read' them back to me. Of course, they weren't really reading but had memorized the text, page by page, and enjoyed being the 'readers.'"

A man named Jonah told me that his son, who has learning difficulties, loved to follow along in his own copy of whatever book his dad was reading. "My son has always been a very small-picture guy, focused on detail, not summary," the father said.

He has always lacked the sort of reasoning it takes to see the big picture: the theme, the connections, all the things that are supposed to be the purpose of literature. But he takes joy in learning and remembering the most minute details.

When he was about eight, we began reading the Harry Potter series together at bedtime. At his prompting, we bought a British and an American edition each time. I would read the American text out loud, and he would follow carefully along in his British edition until that exciting moment every few pages when "truck" would be "lorry" for him, and he would shout it out. We went through the first few books like that.

My kids never sit still. How am I supposed to read to them?

This is tricky, and there's not a short answer. Some children may seem to signal boredom and indifference when in fact they are wide awake and paying attention. That was the case with Gabe Rommely, who has autism. Other kids find it hard to focus, or dislike being confined on a lap. They may wrest themselves away from a restraining arm. These children may only be able to enjoy the reading experience if they're given space to move. For a grown-up this can feel like rejection, which is unfortunate.

So it's worth remembering that there is no "correct" way to read aloud. Given the amazing benefits, we need to make accommodations for all kinds of personalities. Children are people too, as someone said. They vary in the kinds of stories they enjoy, in their desire to be held or touched, and in the amount of landscape description they can tolerate (Molly, Paris, and Violet loved every word of C. S. Lewis's *Prince Caspian*, and sat riveted beside me through every chapter, whereas the minute we got to a windy passage about rivers or cliffs or gorges, Phoebe would trundle off to play with toys on the floor).

A more delicate decision arises when a child seems to prefer the screen to any other company or entertainment. Families have to make their own decisions, but I side with the developmental psychologists who argue that children who spend the bulk of their time with machines are the ones who probably most need a regular whiff of that neurochemical bouquet.

I feel bad. We *used* to read together. I don't know why we stopped. And my kids are still pretty young . . .

Never mind. You can reclaim it! One mother, Amelia DePaola, read aloud to her daughter until the girl was eight and able to read well independently. Several years on, the girl wasn't reading much, and the mother-daughter relationship had become fraught. There was a lot of arguing. The two of them seemed trapped in a cycle of never-ending clashes. When someone sent DePaola a newspaper article (mine, as it happens) about the joy of reading aloud, she was reminded, with a jolt, of how happy those times had been. She proposed to her daughter that they start reading again in the evenings, and to her delight, the girl agreed.

"I ask myself, why did I stop reading aloud?" DePaola told me, when we met for coffee. "Now that I'm reading to her, she's walking around with a book again and I can't believe it, it's like a miracle!"

There was a pause. DePaola's face crumpled, and she had to blot her eyes.

"When I started reading to her aloud again, I felt like I could *rewind time*. She and I had been"—she made fists and knocked them together, to show how they had been fighting—"and I felt like, what else can I do, for us to find a happy ground and not *be* like this all day?

"It really has helped us. At least we have those moments, those magical bonding moments, mother-daughter love. We're in that story together, you know?"

Recapturing the old magic may not always be possible. Lauri Hornik, who had stopped reading when her daughter was ten, tried again when Ruby was thirteen. Hornik described the conversation to me: "I said to her, 'I was thinking, maybe I could start reading to you again and that would be a just nice winding-down-the-evening time for the two of us. What do you think?' And she"—Hornik broke off, laughing—"well, she rejected it entirely."

The moral of the story would seem to be: Once you start, keep going. In a quaint volume from 1907, *Fingerposts to Children's Reading*, Walter Taylor Field has a definitive answer to the question of when parents should stop reading aloud: "At no point whatever."

That's bonkers.

No, no! His point is that we should let *children* decide when to bring the experience to an end, not call it off ourselves. When children were interviewed for Scholastic's 2016 survey of family reading habits, the vast majority said that what they liked best about being read to was that it was "a special time with their parents." As long as they want us to keep reading, why wouldn't we want to keep the emotional connection?

In psychology, there's a concept of human motivation known as self-determination theory. According to this idea, people need to satisfy three intrinsic needs to feel happy and fulfilled. Sebastian Junger summarizes it in his short and excellent book about alienation and connection, *Tribe*. Human beings, he writes, "need to feel competent at what they do; they need to feel authentic in their lives; and they need to feel connected to others."

That passage helped me to understand why reading aloud is such a force for good in family life. The listener gets a great deal out of the experience, as we have seen, but so does the person holding the book. Sociable reading at home helps to satisfy those three intrinsic needs. The more we read aloud, the better we get at it: that's competence. Being present in the moment and giving the people we love our full and openhearted attention: that's authenticity. Building a shared imaginative library of stories, characters, words, and funny lines with children, spouses, or parents: that's connection.

No man is a hero to his valet, the English say. There is a corollary for parents that we may only rarely wish to admit: our children look up to us when they are small and do not suspect what we know to be true about ourselves; that even as we may strive to be fair and wise and conscientious, we blow it sometimes. We are flawed and fallible and not always perfect in temper. When children are approaching adolescence, they start to notice this, and it can be an uncomfortable transition for everyone involved.

I might seem to be straying from the subject at hand, but for me it all ties together. Reading every day with children can't guarantee perfect outcomes for any family—not in grades, not in happiness, not in relationships. But it is as close to a miracle product as we can buy, and it doesn't cost a nickel. As a flawed, fallible person with an imperfect temper, I know that reading every night is not just the nicest thing I've done with my children but represents, without question, the best I have been able to give them as their mother.

If we do what you recommend, and read to our kids every day, how soon will we see a difference? The vocabulary growth, the enhanced attention spans, all the rest. How long does it take?

Results may vary, as disclaimers say, but I hope we can draw some conclusions from what happened in the Rashid household during the three-month trial. I suspect the outcome will impress you. I'm a true believer, and I was amazed.

*   *   *

THE BIG-SCREEN TV was off when I rang the front-door bell at the end of the summer. When Julie Rashid let me in, there was a subtle change in the atmosphere. The family was just as cheerful as when I'd visited the last time, but now there was a noticeable feeling of calm. I entered the family room, greeted Alex, and found the place strewn with books, as it had been three months before—but this time it was because the children had spent the last half hour hunting around the house to retrieve the books I'd loaned them.

What a change! The baby, now two, was no longer stomping around, sliding on books and crushing their spines. Having come to the door when I arrived, he trotted after me into the family room, stood for a moment in the middle of the array, and then squatted down by a board-book edition of *Go, Dog. Go!* and began turning its pages.

"Wow," I said, looking at his parents for confirmation.

"Ethan used to chew on books," said Julie, "but in the past two months he's started to flip through them instead."

"I wonder why," I said, and we all laughed.

"It's amazing how the children changed," Julie said. "*He* turned from chewing to flipping," she said, pointing to the baby,

and with Joseph, all of a sudden vocabulary words were there! Eva just adored reading, and actually took over the reading role at one point.

It was something we'd look forward to. We even invited an aunt to join us, we invited my mother to read—we made it an activity. We'd just decide: what *time* did we want to read, not *if* we wanted to read. Sometimes it was 9:00 a.m., sometimes it was 4:00 p.m. A lot of decisions depended on what the baby had done during the day. Did he get up at 5:30? Did he nap? So a lot of our reading-time decisions depended on Ethan.

And we liked it during the day, it was fun—if it was a rainy day or a hot day, it was nice to settle down at two o'clock and read. And I would say we averaged forty-five minutes to an hour every time.

Reading in the bath had been a huge success. "It was fabulous," Julie said. "They're contained, they're listening. I would get through at least six books while they played. If I started skipping, because I wanted to get through, because the baby was done with the bath or something, I'd get, you know, 'You missed that line, you have to go back!'"

Joseph wriggled with pride on the leather sofa. On the floor, his diaper peeping out, Ethan remained engrossed in P. D. Eastman's pictures of a dog party.

"Joseph has memorized a number of books, like *Brush of the Gods*—"

"And *Finding Winnie*," Eva put in.

"And *The Bear Ate Your Sandwich*," said Julie.

"—*Demolition*, and *King Arthur's Very Great Grandson*—"

"I'd say after the first month everyone had their requests," Julie said. "We'd read a new story, but there were always at least two or three please-Mommy-could-we-read-again books."

Still crouched on the floor, Ethan had moved on to *Good Night, Gorilla* and was slowly paging through the color-saturated pictures of animals slipping one by one out of their cages and forming a silent train behind the oblivious zookeeper. Julie noticed me looking at him. "This never happened three months ago," she said.

Eva spoke up: "We didn't read a lot of chapter books, but most of the picture books we read, I'd say we read about a hundred times."

"And he got the most out of it, he's much better in the mornings," said Alex, indicating Joseph, who was now absorbed in the pages of *My Father's Dragon*. "It was good seeing him even at nighttime being able to sit down and focus, and listen, and ask questions."

Alex laughed. "It was cool, it was what we all need—especially now, there's too much screen time in this house. And that's something that I can't avoid myself. I'm probably the worst at it, because I'm constantly researching on the Internet. Yeah, you're reading but you're still in front of the screen, which isn't the best role model for the three of them."

He picked up a paperback of Mordecai Richler's *Jacob Two-Two*

Meets the Hooded Fang and shook it for emphasis. "It's so different, the touch and feel of books."

In the end, Alex had not been a big participant in the reading time, but he loved the effects. "I see a lot of benefits for both the boys," he said, "a tremendous amount."

Julie broke in: "And you could tell the difference, which was amazing."

"I mean, his *vocabulary*," Alex said, tipping his head toward Joseph again.

"The vocabulary that came out of his mouth!" Julie agreed. "And I would remember it was from a certain story! It was amazing to me. He'd use a word properly in the context and I'd say, 'Where did he hear that word?' and I'd be like, 'Oh, it was that book . . .'"

And I thought: *Bingo.*

AFTERWORD

In *The Little Prince*, a desert fox confides a secret to the small visitor from Asteroid B-612, which he in turn tells the author, Antoine de Saint-Exupéry. It has become the most famous quotation from that famous book: "It is only with the heart that one can see rightly; what is essential is invisible to the eye."

Walking in the bleak, beautiful desert with the little prince, Saint-Exupéry is "astonished by a sudden understanding." He tells us:

> When I was a little boy I lived in an old house, and legend told us that a treasure was buried there. To be sure, no one had ever known how to find it; perhaps no one had even looked for it. But it cast an enchantment over that house. My home was hiding a secret in the depths of its heart. . . .
>
> "Yes," I said to the little prince. "The house, the stars, the desert—what gives them their beauty is something that is invisible!"

What is essential is invisible to the eye.

It seems to me that the promise and treasure of reading aloud is a lot like that. As spectacle, it is dull. There is a grown-up sitting with a child or two, or perhaps with a half-dozen other grown-ups at two round tables. There is a book, or a stack of them, or a sheaf of photocopied poems. There is a clock, and a bit of time. There is the human voice, reading; there are human ears, listening.

What makes the experience beautiful, and essential—the richness of the emotional exchange, the kindling of the mind, the voyaging in imagination, the sharing of culture and pathos and humor—is invisible to the eye.

Yet its effects can be seen. And they are lovely.

We live in a time of immense complexity, dizzying and dazzling sophistication that would seem to make a mockery of simpler ways and things. Yet there is magic in simplicity. Flour and water and yeast make bread. Pen and paper and imagination make a portrait, a landscape, a novel. Two people and a book together make an experience of force and significance out of all proportion to the time it takes.

When the writer and illustrator Anna Dewdney knew that brain cancer was taking her early from this world, she asked that, in lieu of a funeral or memorial service, her friends and the people who loved her books would read to a child. She knew what was essential. In a piece for the *Wall Street Journal's Speakeasy* blog, she wrote: "When we read with a child, we are doing so much more than teaching him to read or instilling in her a love of language. We are doing something that I believe is just as powerful, and it is something that we are losing as a culture: by reading with a child, we are teaching that child to be human."

Reading aloud is a small thing, yet profound. To read to someone you love is the simplest of gifts, and one of the greatest. All that is required for a long, happy string of enchanted hours is for someone to take the trouble to make it happen.

Surely that is something we can aspire to do. With love.

ACKNOWLEDGMENTS

Isaac Newton usually gets credit for coining an expression that captures the cumulative nature of knowledge. In a 1676 letter, the father of modern physics wrote: "If I have seen further it is by standing on ye shoulders of Giants." In fact, he wasn't the first to use the phrase. True to the metaphor, he got it by climbing onto the shoulders of earlier thinkers who had said much the same thing.

It is encouraging to have at least one thing in common with Isaac Newton! To write this book I have, like him, clambered around on the shoulders of giants, jumping from one to another to see vistas that would never have been available to me had I stayed on the ground, locked in the single perspective of my own eyes.

Not that personal experience doesn't count. In many years of reading aloud to my children, I knew that something big was happening. I could feel it. The mutual engagement, the speed and breadth of their language acquisition, the shared vicarious thrills and laughs over the stories we read; the experience was both practical and mystical. Yet without ye Giants—the researchers, doctors, professors, librarians, writers, and philanthropists who have devoted their energies to exploring the hidden secrets of reading out loud—I could not have told you *why* it was happening.

I have debts of gratitude to a great number of people. I am so thankful for my grandparents, Mary and Frank Gillman and Barbara and Allan Cox, who made my parents possible, and who read to them (and to me). I am grateful for the loyal and unstinting love of my mother, Noel Cox, and of my father and stepmother, Allan Cox and Grace Simonson.

Lisa Wolfinger, thank you for showing me how to put reading aloud at the heart of family life. Thank you, Robert Messenger, for giving "The Great Gift of Reading Aloud" such a sensitive and demanding edit and such great acreage in the *Wall Street Journal* in the summer of 2015. Thank you, Mary Ortiz, for steering me in the right direction afterward, and giving me a push.

Thank you, Stephen Barbara, my amazing agent, for your wise counsel and buoyant good humor as draft succeeded draft. I'm grateful, too, to the rest of the team at Inkwell Management, in particular Lyndsey Blessing and Emma Gougeon.

Thank you, Gail Winston, my wonderful editor at Harper, who has the merciful acupuncturist's gift of knowing exactly where to stick the needle.

Thank you to the dazzling Emily Taylor, especially for herding the editorial cats, and to Robin Bilardello and Fritz Metsch for the beautiful cover art and interior design. Thank you, Miranda Ottewell, for saving me from typos and snafus.

For editorial tough love and the generous gift of their time through numerous rereadings, I am deeply indebted to Mona Charen, Danielle Frum, David Frum, Rosemary Wells, and Diane Zeleny.

For inspiration, anecdotes, and expertise that made this book possible, I am so grateful to (in alphabetical order, because otherwise it's impossible) Dr. Mohammed Kabir Abubakar, Anne Applebaum, Lizzie Atkinson, Dr. Barbara Bean, Claudia Zoe Bedrick, Patrick Braillard, Stuie Brown, Morten Christiansen, Lora Coonce, Dan Coupland, Melissa Davidson, Carl Dennis, Maureen Ferguson, Jane Fidler, Luke Fischer, Amy Freeman, Kate Fulton, Sally Gannon, Reuel Gerecht, Roberta Michnick Golinkoff, Kelli Gray, Chen Guangcheng, Amy Guglielmo, Paul Higgins, Dr. Scott Holland, Annie Holmquist, Lauri Hornik, Dr. Tzipi Horowitz-Kraus,

Dr. John Hutton, Dr. Candace Kendle, Dr. Perri Klass, Deborah Lancman, Jamie Lingwood, Alyson Lundell, Matt Mehan, Taylor Monaco, Christine Nelson, Walter Olson, Marshall Peters, Susan Pinker, Steve Pippin, Andrew Pudewa, Christine Rosen, Caroline Rowland, Matthew Rubery, Laura Amy Schlitz, Roger Scruton, Dr. Suna Seo, Christina Hoff Sommers, Dr. Siva Subramanian, Catherine Tamis-LeMonda, Maria Tatar, Puffin Travers, Jack Wang, Victoria Wells, Natasha Whitling, Marianne Worley, and Paul O. Zelinsky.

I'm particularly grateful to members of the Barsantini, Baylinson, Carroccio, Daniels, DeMuth, Duggan, Grey, Mullner, Nader, Reese, Rossiter, Sikorski, and Yeager families, who know who they are, as well as Chris Carduff at the *Wall Street Journal*, and to the staff of the Story Museum in Oxford, England. I want also to thank Gwendolyn van Paasschen for making the elegant home she shared with John Makin my hermitage during an initial crunch-stage of writing, and Drs. Toby and Liz Cox for their Utah hospitality and for sharing their knowledge of pediatrics. Thank you, too, Erich Eichman, my longtime editor at the *Wall Street Journal* and one of the world's true gentlemen, for your goodness and forbearance over many years. In a collaboration of more recent vintage, I am so thankful for the spirited and enterprising Miranda Frum at Kool-Haus Media. Any "enchantment" you see online is undoubtedly her doing.

In a broader spirit of gratitude, I'd like to salute certain individuals who have made reading aloud their cause. These champions include Claudia Aristy, whose energetic stewardship of the Reach Out and Read program at Bellevue Hospital in New York brings this achievable miracle into the lives of countless low-income kids. Hurrah too for Katrina Morse and her colleagues at the Family Reading Partnership in Ithaca, New York, who have dotted the landscape of Tomkins County with cheerful exhortatory posters and little red bookshelves stocked with freebies so that families can help themselves. Hats off to Lester Laminack and Mem Fox, two especially eloquent and enthusiastic advocates of reading aloud; to Dolly Parton, whose philanthropy through her charity, Imagination Library, has put an incredible 100 million plus books into the hands of children; to the founders of Reach Out and Read, Dr. Robert Needlman, Dr. Barry Zuckerman, and Dr. Alan Mendelsohn; to Sarah Mackenzie, the creator of the *Read-Aloud Revival* podcast; and to Jim Trelease, author of *The Read-Aloud Handbook*, who since 1982 has perhaps done more than any other person to spread the gospel of reading aloud in the United States.

My life as a writer and mother is inconceivable without the love and support of my husband, Hugo Gurdon, my companion in a thousand swashbuckling adventures. And the Chogen? Well, obviously, none of this would have been possible without you. Molly, Paris, Violet, Phoebe, and Flora, reading to you has been the greatest privilege of my life. Being your mother is the greatest joy.

NOTES

·—◆·◆·◆—·

Introduction

xiii "We let down our guard": Kate DiCamillo, email exchange with the author, early 2015, quoted in Meghan Cox Gurdon, "The Great Gift of Reading Aloud," *Wall Street Journal*, July 10, 2015.

xiii "the big disconnect": Catherine Steiner-Adair, *The Big Disconnect: Protecting Childhood and Family Relationships in the Digital Age* (New York: Harper, 2013).

xv "the cotton wool of daily life": Virginia Woolf, *Moments of Being*, quoted on Goodreads, https://www.goodreads.com/work/quotes/900708-moments-of-being-autobiographical-writings.

xix If I were Glinda: In L. Frank Baum's 1900 classic, *The Wonderful Wizard of Oz*, Glinda achieves her benevolent effects without a wand. Since most readers will know Glinda best from the 1939 MGM movie, in which she has a wand, I'm taking the liberty of giving her one here, too.

Chapter 1: What Reading to Children Does to Their Brains

1 In 1947: https://www.dior.com/couture/en_us/the-house-of-dior/the-story-of-dior/the-new-look-revolution; https://www.history.com/this-day-in-history/jackie-robinson-breaks-color-barrier; https://en.wikipedia.org/wiki/Goodnight_Moon

1 a gazillion copies: As of April 12, 2018, the book has sold an estimated 48 million copies, according to Wikipedia.

3 "*Goodnight Moon* time": Robert D. Putnam, *Our Kids: The American Dream in Crisis* (New York: Simon & Schuster, 2015), 126.

3 a certain chilled enclosure: MRI rooms located down the hall from the Reading and Literacy Discovery Center, Cincinnati Children's Hospital Medical Center, visited by author February 7–8, 2017.

4 This bed is narrow: These descriptions derive from the author's own experience of undergoing a reading experiment using fMRI to help test a protocol at Cincinnati Children's. Author has the brain scan to prove it!

4 Half a dozen miles away: Blue Manatee Bookstore, Cincinnati, visited by author on February 7, 2017.

7 A human brain was floating: Dr. John S. Hutton, interviews by author, February 7–8, 2017, Cincinnati Children's Hospital.

7 an exciting study: John S. Hutton et al., "Home Reading Environment and Brain Activation in Preschool Children Listening to Stories," *Pediatrics* 136, no. 3 (2015): 466–78.

8 "They come in the morning": Andrea Roure, preschool teacher, National Child Research Center, Washington, DC, interview by author, July 20, 2017.

9 two additional first-of-their-kind fMRI-based papers: John S. Hutton et al., "Story Time Turbocharger? Child Engagement During Shared Reading and Cerebellar Activation and Connectivity in Preschool-Age Children Listening to Stories," *PLoS One* 12, no. 5 (2017): e0177398; Tzipi Horowitz-Kraus and John S. Hutton, "Brain Connectivity in Children Is Increased by the Time They Spend Reading Books and Decreased by the Length of Exposure to Screen-Based Media," *Acta Paediatrica* 107, no. 4 (April 2018), https://doi.org/10.1111/apa.14176.

9 greater activation in their cerebellum: Hutton et al., "Story Time Turbocharger."

9 a time of such intense formation: For a useful description of brain growth in infancy and early childhood, see "Early Childhood Development: The Key to a Full and Productive Life," Unicef, https://www.unicef.org/dprk/ecd.pdf.

9 "Reading regularly with young children": Council on Early Childhood, "Literacy Promotion: An Essential Component of Primary Care Pediatric Practice," *Pediatrics*, June 23, 2014, https://doi.org/10.1542/peds.2014–1384.

10 almost half of young children: A. R. Lauricella et al., *The Common Sense Census: Plugged-in Parents of Tweens and Teens* (San Francisco: Common Sense Media, 2016), 13.

10 Children ages eight and under: Jacqueline Howard, CNN, "Kids Under 9 Spend More Than 2 Hours a Day on Screens, Report Shows," October 19, 2017, http://www.cnn.com/2017/10/19/health/children-smartphone-tablet-use-report/index.html.

10 Older children are even more absorbed: Vicky Rideout, Sita Pai, "U.S. Teens Use an Average of Nine Hours of Media Per Day, Tweens Use Six Hours," Common Sense Media, November 3, 2015, 13. There is a distinction between young people's use of screen-based media and consumption of all entertainment media (which would include activities such as reading and listening to music). The report, as evidenced by the title, emphasizes the longer span of time devoted to total media use. For the

purposes of my argument, I am noting the time that teens spend using screens, specifically.

11 Another first-of-its-kind study: John S. Hutton et al., "Shared Reading Quality and Brain Activation During Story Listening in Preschool-Age Children" *Pediatrics,* December 2017, https://doi.org/10.1016/j.jpeds.2017 .08.037.

12 "You're getting a little bit cooking": Hutton, interview.

14 the Goldilocks effect: John S. Hutton et al., "Goldilocks Effect? Illus-trated Story Format Seems 'Just Right' and Animation 'Too Hot' for Integration of Functional Brain Networks in Preschool-Age Children," https://www.eurekalert.org/pub_releases/2018–05/pas-nsm042618.php.

14 What children don't get from one: Horowitz-Kraus, "Brain Connectiv-ity in Children."

15 "unfairly disadvantaging": Adam Swift, interview by Joe Gelonesi, "Is Hav-ing a Loving Family an Unfair Advantage?," *Philosopher's Zone,* ABC, May 1, 2015, http://www.abc.net.au/radionational/programs/philosopherszone /new-family-values/6437058.

15 Swift was using the phrase: Adam Swift, email exchange with author, May 2015.

15 calling *"Goodnight Moon* time": Putnam, *Our Kids,* 126–27.

15 "differences in parenting": Putnam, 123.

16 A 2012 study: "The First Eight Years: Giving Kids a Foundation for Life-time Success," Annie E. Casey Foundation, November 2013. Cited in (among others) Michael Alison Chandler, "Children from Poor Families Lag in Cognitive Development and Other Areas, Report Says," *Washing-ton Post,* November 3, 2013. See also, Putnam, *Our Kids,* 127.

16 the word gap: This famous phrase entered the academic lexicon with the publication of C. Hart and T. Risley, *Meaningful Differences in the Everyday Experience of Young American Children* (Baltimore: Brookes, 1995).

16 A 2017 study put the number: Jill Gilkerson et al., "Mapping the Early Language Environment Using All-Day Recordings and Automated Analysis," *American Journal of Speech-Language Pathology* 26 (May 2017), https://doi.org/10.1044/2016_AJSLP-15–0169.

16 two classroom subjects might seem on the surface to have little in com-mon: Consider the commonly held notion that one is either a "math person" or a "language person," or perhaps just "better at math." In America, at least, we tend to imagine that any given person's talent leans toward one or the other. In fact, the two disciplines have import-ant qualities in common, not least the skills we need to understand them.

16 "If you can't do fifth-grade reading problems": Candace Kendle, phone interview by author, late February 2016. John Hutton is the "spokes-doctor" of Kendle's campaign, Read Aloud 15 MINUTES, http://read aloud.org/.

17 As CEO of a clinical research organization: In 1981 Kendle cofounded Kendle International, a clinical research and drug development company, serving as its chief executive officer from 1981 to 2011.

17 The numbers may be worse: 2015 Mathematics & Reading Assessments, The Nation's Report Card, https://www.nationsreportcard.gov/reading _math_2015/#reading?grade=4.

17 Something like 20 percent: The statistics in this paragraph are widely quoted and have their origins in an April 2002 report, the National Assessment of Adult Literacy, conducted by the National Center for Educational Statistics (https://nces.ed.gov/pubs93/93275.pdf), and were confirmed by a 2013 international survey published by the Organization for Economic Cooperation and Development, cited here: https://www.insidehighered.com/news/2013/10/08/us-adults-rank-below-average-global-survey-basic-education-skills.

17 "We could narrow the achievement gap": Rosemary Wells, email exchange with the author, August 2017.

18 56 percent of families reported: *Kids & Family Reading Report*, 6th ed. (New York: Scholastic, 2017), http://www.scholastic.com/readingreport /reading-aloud.htm.

18 the numbers are actually dropping: Alison Flood, "Only Half of Pre-School Children Being Read to Daily, UK Study Finds," *Guardian*, February 21, 2018, https://www.theguardian.com/books/2018/feb/21/only -half-of-pre-school-children-being-read-to-daily-study-finds.

19 "Any time, any place": The Family Reading Partnership, Ithaca, NY, www.familyreading.org.

19 "Though books were scarce": Roger McGough, quoted in Antonia Fraser, ed., *The Pleasure of Reading* (London: Bloomsbury, 1992), 138. Cited in Maria Tatar, *Enchanted Hunters: The Power of Stories in Childhood* (New York: W. W. Norton, 2009), 226.

Chapter 2: Where It All Began: Once Upon a Time in the Ancient World

21 a glossy black-and-ocher amphora: It can be seen on the British Museum website at http://www.britishmuseum.org/research/collection_online /collection_object_details.aspx?objectId=399287&partId=1.

21 "stitcher of songs": For a brief description of this term, and the role of the rhapsode, see http://www.oxfordreference.com/view/10.1093/oi /authority.20110803100418375.

22 the sheer size of the thing: My paperback edition of Homer's *Iliad*, translated by Robert Fagles (New York: Penguin, 1998), weighs in at two pounds exactly, and it's two inches thick. A hefty tome indeed!

22 recite onward without a hitch: Gregory Nagy, *The Ancient Greek Hero in 24 Hours* (Cambridge, MA: Belknap Press, 2013), 246–47.

23 "the liquid tapestry": This wonderful expression appears in Rushdie's *Haroun and the Sea of Stories* (New York: Penguin, 1991). For storytelling as one of the great human universals, see Donald Brown, *The Human Universals* (Philadelphia: Temple University Press, 1991).

23 So far as we can tell: We see prehistoric cave paintings in Sister Wendy and Patricia Wright, *Sister Wendy's Story of Painting* (New York: DK, 1994).

23 *Gilgamesh*: https://www.ancient.eu/gilgamesh/.

23 *Mahabharata*: https://www.britannica.com/topic/Mahabharata.

23 *Ramayana*: https://www.britannica.com/topic/Ramayana-Indian-epic.

23 *Beowulf*: https://www.britannica.com/topic/Beowulf.

23 *Völsunga* saga: https://www.britannica.com/topic/Volsunga-saga.

23 *Sundiata*: https://en.wikipedia.org/wiki/Epic_of_Sundiata.

23 *Mabinogion*: https://www.britannica.com/topic/Mabinogion.

24 *The Thousand and One Nights*: https://www.britannica.com/topic/The -Thousand-and-One-Nights.

24 *Kalevala*: https://www.britannica.com/topic/Kalevala.

24 Long before Johannes Gutenberg: Alberto Manguel, *A History of Reading* (New York: Penguin, 2014), 59–60.

24 To read at all was to read out loud: Manguel, 43, 45.

24 the "primordial" languages of the Bible: Manguel, 45.

25 "Much merit is supposed to be derived": Joseph Edwin Patfield, *The Hindu at Home: Being Sketches of Hindu Daily Life* (Ann Arbor, MI: Society for Promoting Christian Knowledge, 1896), 178.

25 Plutarch writes of the way that Alexander: Manguel, *History of Reading*, 43.

25 the old man's peculiar technique: Augustine quoted in Manguel, 43.

25 "For Augustine": Manguel, 45.

25 Yet as Dante observed: Manguel, 251.

26 about 14 percent of the world's adult population: UNESCO Institute for Statistics, http://uis.unesco.org/sites/default/files/documents/fs45 -literacy-rates-continue-rise-generation-to-next-en-2017_0.pdf.

26 "It was . . . and it was not": Maria Tatar, phone interview by author, March 26, 2017.

27 "Who can catalog the myriad ways": Laura Miller, *The Magician's Book: A Skeptic's Adventures in Narnia* (New York: Little, Brown, 2008), 284.

27 One autumn afternoon not so long ago: Author visit to the Bard Competition, October 30, 2015, Heights School, Potomac, MD.

29 "After crying a good deal": Charles Dickens, *The Letters of Charles Dickens* (New York: Macmillan, 1893), 463.

29 "a remarkably good fellow": Dickens, *Letters*, 463.

29 His readers in the United States: This celebrated incident is recorded, among other places, by Richard Lederer, "Remembering the Great Charles Dickens," *Language Magazine,* https://www.languagemagazine.com/the-great-charles-dickens/.

30 In the time of Dickens: Abigail Williams, *The Social Life of Books: Reading Together in the Eighteenth-Century Home* (New Haven: Yale University Press, 2017), 77.

30 "People shared their literature": Williams, 10.

30 From a commonplace diversion: Williams, 14–15, 34, 147.

30 "like an ignorant Boy": Williams, 84.

30 "Though she perfectly understands the characters": Patricia Howell Michaelson, "Reading *Pride and Prejudice,*" *Eighteenth-Century Fiction* 3 (October 1990): 65–76, http://ecf.humanities.mcmaster.ca/3_1michaelson/.

30 "I would ask that you imagine": Manguel, *History of Reading*, 257.

31 "Cheerful . . . Stern . . . Pathos": Manguel, 257–58.

31 While Dickens was entertaining audiences: Dickens famously undertook grueling book tours between the years 1858 and 1867, including a lucrative sojourn to the United States. The author writes about his experiences in Dickens, *Letters*. See also Matt Shinn, "Stage Frights," *Guardian,* January 30, 2004, https://www.theguardian.com/stage/2004/jan/31/theatre.classics.

31 "We were never able to be sure": Christopher W. Czajka, "How the West Was Fun: Recreation and Leisure Time on the Frontier," Frontier House, PBS, https://www.thirteen.org/wnet/frontierhouse/frontierlife/essay9.html.

32 "Come girls": Laura Ingalls Wilder, *The Long Winter*, illustrated by Garth Williams (New York: Harper Trophy, 1994), 171.

32 "After a moment Mary said": Wilder, 174–75.

33 "'You girls choose a story'": Wilder, 184–85.

33 In 1865, as Alberto Manguel recounts: Manguel, *History of Reading*, 110–14.

34 "In the mornings, he read": Ann L. Henderson, Gary R. Mormino, and Carols J. Cano, eds., *Spanish Pathways in Florida/Los Caminos Espanioles en la Florida, 1492–1992* (Sarasota, FL: Pineapple Press, 1992), 284.

34 *The Count of Monte Cristo*: Manguel, *History of Reading*, 113.

34 An 1873 magazine sketch: From the *Practical Magazine*, New York, reprinted in Manguel, 112.

34 first recording of the human voice: For a brief description of the incident, and Edison's view of it, see "Edison Reading 'Mary Had a Little Lamb,'" *Public Domain Review*, http://publicdomainreview.org/collections/edison -reading-mary-had-a-little-lamb-1927/.

34 The original phonograph track: Lisa Brenner, "Hear Thomas Edison's Earliest Known Recording from 1878 for the First Time (Audio)," KPCC radio website, October 25, 2012, http://www.scpr.org/blogs /news/2012/10/25/10712/hear-thomas-edison-sing-rare-1878-audio -restored-f/.

35 These readers were inhuman: Matthew Rubery, interview by author, London, June 22, 2016.

35 "There was a real sense": Matthew Rubery, *The Untold Story of the Talking Book* (Cambridge, MA: Harvard University Press, 2016).

35 The first full-length recorded books: Matthew Rubery, "Another Historic Talking Book Found," *Audiobook History* (blog), November 28, 2016, https://audiobookhistory.wordpress.com/author/mattrubery/.

35 For the war-blind: Rubery, interview.

36 Not until the mid-1970s: Matthew Rubery, "What Is the History of Audiobooks?," symposium discussion excerpted on Center for the History of the Book, University of Edinburgh, August 21, 2015, https://www .ed.ac.uk/literatures-languages-cultures/chb/books-and-new-media /dr-matthew-rubery.

36 Audiobook enthusiasts were inclined: Rubery told me that he became interested in the topic of audiobooks when a distinctly nonacademic friend of his father's "got very excited and wanted to tell me he'd read a book. And then he got apologetic and said, Well, actually I didn't read it, I listened to it. . . . It didn't matter to me, but I did think about that apology and noticed it everywhere after that. And I wanted to get to the roots of that curious shame. Why would the experience of listening to a book be treated differently than the experience of reading a book? It's viewed as cheating all the time." Rubery, interview.

36 a $3.5 billion industry: Michael Kozlowski, "Global Audiobook Trends and Statistics for 2017," *Goodereader.com*, December 18, 2016, https:// goodereader.com/blog/digital-publishing/audiobook-trends-and-statistics -for-2017. See also Tom Webster, "Monthly Podcast Consumption Surges to More Than One in Five Americans," *Edison Research*, March 7, 2016, http://www.edisonresearch.com/monthly-podcast-consumption -surges-to-more-than-one-in-five-americans/.

36 the shift in acceptability: Kevin Roose, "What's Behind the Great Pod-cast Renaissance?," *New York* magazine, October 30, 2014, http://nymag .com/daily/intelligencer/2014/10/whats-behind-the-great-podcast -renaissance.html.

36 It was a sign: Rubery, interview.

36 a hipster resurgence: Calvin Reid, "HarperAudio Goes Retro with New Vinyl Audiobook Series," *Publishers Weekly*, January 18, 2018, https:// www.publishersweekly.com/pw/by-topic/industry-news/audio-books /article/75843-harperaudio-goes-retro-with-new-vinyl-audiobook-series .html.

36 Most new cars now come fitted: Roose, "What's Behind the Great Pod-cast Renaissance?"

38 "People who like audiobooks often try": Rubery, interview.

38 a scene in *The Odyssey*: Homer, *Odyssey*, trans. Robert Fitzgerald, book 9, viewable at http://mbci.mb.ca/site/assets/files/1626/homer_sodyssey.pdf.

39 If we are wise: Maria Tatar, *Beauty and the Beast: Classic Tales About Animal Brides and Grooms from Around the World* (New York: Penguin, 2017), 21.

Chapter 3: Reading Together Strengthens the Bonds of Love

41 Apple tablet had been on the market: Roger Fingas, "A Brief History of the iPad, Apple's Once and Future Tablet," *Apple Insider*, April 3, 2018, https://appleinsider.com/articles/18/04/03/a-brief-history-of-the-ipad -apples-once-and-future-tablet.

41 *Goodnight iPad*: David Milgrim, *Goodnight iPad* (New York: Blue Rider, 2011).

41 Rare is the household: Adam Alter, *Irresistible: The Rise of Addictive Tech-nology and the Business of Keeping Us Hooked* (New York: Penguin, 2017), 13–19.

42 "the latest and most powerful extension": Virginia Heffernan, *Magic and Loss* (New York: Simon & Schuster, 2016), 21.

42 "a massive and collaborative work": Heffernan, 8.

42 Screens have rushed into childhood: Jean Twenge, "Have Smartphones Destroyed a Generation?," *Atlantic*, September 2017, https://www.the atlantic.com/magazine/archive/2017/09/has-the-smartphone-destroyed -a-generation/534198/.

42 According to Jean Twenge: Jean Twenge, "What Might Explain the Unhappiness Epidemic?," *The Conversation*, January 22, 2018, http://the conversation.com/what-might-explain-the-unhappiness-epidemic-90212.

42 "We found that teens": Twenge.

42 We adults, meanwhile: Alter, *Irresistible*, 15.

43 "You are picking up your child!": Lucia I. Suarez Sang, "'Get Off Your Phone!!' Daycare's Message to Parents Goes Viral," Fox News, February 1, 2017, http://www.foxnews.com/us/2017/02/01/get-off-your-phone -day-cares-message-to-parents-goes-viral.html.

44 "technoference": B. T. McDaniel, and J. S. Radesky, "Technoference: Parent Distraction with Technology and Associations with Child Behavior Problems," *Child Development* 89 (2017), https://doi.org/10.1111 /cdev.12822.

45 "I feel like, ughhh, sad": Steiner-Adair, *Big Disconnect*, 13.

45 "Our household is eerily silent": Steiner-Adair, 27.

45 the interactions of thirty middle-class Los Angeles families: Belinda Campos et al., "Opportunity for Interaction? A Naturalistic Observation Study of Dual-Earner Families after Work and School," *Journal of Family Psychology* 23, no. 6 (2009), https://doi.org/10.1037/a0015824. Cited in Susan Pinker, *The Village Effect: How Face-to-Face Contact Can Make Us Healthier, Happier, and Smarter* (New York: Spiegel & Grau, 2014), 167–69.

45 "Despite the mountain of evidence": Susan Pinker, email exchange with author, June 2, 2017.

46 "I could feel her voice": Quoted in Tatar, *Enchanted Hunters*, 231.

46 "a whole bouquet of neurochemicals": Quoted in Alter, *Irresistible*, 228–29.

47 "A tsunami of neurochemical benefits": Pinker, email exchange.

47 "The script is right there": Katrina Morse, acting director, Family Reading Partnership, Ithaca NY, interview by author, April 25, 2016.

47 a team of neuroscientists at Princeton has discovered: Greg J. Stephens, Lauren J. Silbert, and Uri Hasson, "Speaker-Listener Neural Coupling Underlies Successful Communication," *PNAS* 107, no. 32 (August 10, 2010): 14425–30, http://www.pnas.org/content/107/32/14425. Cited in Geoff Colvin, *Humans Are Underrated: What High Achievers Know that Brilliant Machines Never Will* (New York: Portfolio/Penguin, 2016), 152.

47 "Storyteller and hearer are connecting": Colvin, 152.

48 "kind of conspiracy": Mem Fox, *Reading Magic: Why Reading Aloud to Our Children Will Change Their Lives Forever* (Orlando, FL: Harcourt, 2001), 10.

48 "When I got here": Claire Nolan, interview by author, Georgetown University Hospital NICU, April 6, 2017.

49 "We know that parental voice is important": Dr. Mohammed Kabir Abubakar, interview by author, April 25, 2016.

49 Writing about a 2011 experiment there: Maude Beauchemin et al., "Mother and Stranger: An Electrophysiological Study of Voice Processing in Newborns," *Cerebral Cortex* 21, no. 8 (2011), https://doi.org/10.1093 /cercor/bhq242. Cited in Pinker, *Village Effect*, 127.

49 "the language circuitry in a newborn's brain": Pinker, 127.

49 The earlier a child arrives: Dr. K. N. Siva Subramanian, Georgetown University Hospital, interview by author, April 28, 2016.

49 A baby born at even thirty-six weeks: Jennifer E. McGowan et al., "Early Childhood Development of Late-Preterm Infants: A Systematic Review," *Pediatrics* 127, no. 6 (2011), https://doi.org/10.1542/peds/2010.2257.

50 in concert with the human voice: "Even premature infants are sensitive to social context and will vocalize in the neonatal intensive care unit significantly more when a parent is present." Roberta Michnick Golinkoff et al., "(Baby) Talk to Me: The Social Context of Infant-Directed Speech and Its Effects on Early Language Acquisition," *Current Directions in Psychological Science* 24, no. 5 (2015), https://doi.org/10.1177/0963721415595345.

50 In the spring of 2017: Details and findings of the Georgetown University Hospital investigation in Siva, interview; Dr. Suna Seo and Abubakar, interviews.

50 "Inside the incubator it gets really humid": Seo, interview.

51 The doctors and nurses at Georgetown: Abubakar, interview.

51 Their voices are a kind of curative: Pinker, *Village Effect*, 126–27.

52 "Reading promotes better interaction": Abubakar, interview.

52 The story of one Georgetown NICU patient: Lori Green, conversation with the author, December 3, 2015, details confirmed in email exchanges, February 15–21, 2017.

53 the healing analgesic: This phrase is gratefully adapted from one used by Susan Pinker, "homegrown analgesic for babies," to describe the effects of skin-to-skin bonding, or kangaroo care. Pinker, *Village Effect*, 130.

53 There's no retrieving those files: This phenomenon is known as "childhood amnesia." A short explanation can be found here: Janice Wood, "What's Your Earliest Memory?" *Psychcentral.com*, https://psychcentral.com/news/2014/01/26/whats-your-earliest-memory/64982.html.

54 "One of the unexpected joys": Bruce Handy, *Wild Things: The Joy of Reading Children's Literature as an Adult* (New York: Simon & Schuster, 2017), xvi–xvii.

54 "Aside from the immediate pleasure": Handy, 265.

55 One woman told me how: Tayla Burney, quoted in Gurdon, "Great Gift."

55 "She also had a box": Beatrice Frum, conversation with the author, 2015.

56 Tolkien describes the possessions: J. R. R. Tolkien, *The Fellowship of the Ring* (New York: Mariner/HMH, 1994), 36–37.

58 Marine Corps commandant Robert Neller: Remarks at the Fourth Annual Tribute to Military Families under the auspices of United Through Reading, Washington, DC, May 24, 2017.

58 was the first to track: Sarah O. Meadows et al., "The Deployment Life Study: Longitudinal Analysis of Military Families Across the Deployment Cycle," Rand Corporation, 2016, https://doi.org/10.7249/RR1388.

59 "Jack, my youngest": Alice Kirke, phone interview by author, November 11, 2016; interview by author, May 24, 2017; email exchanges, late May/early June 2017. Also Kevin Kirke, interview by author, May 24, 2017.

60 The military charity that facilitated: Kenneth Miller, "A Soldier's Last Bedtime Story," *Reader's Digest*, March 2017, 83.

60 "Every time they got new books": Taylor Monaco, phone interview by author, November 11, 2016.

61 In 2017, UTR surveyed": United Through Reading Beneficiary Surveys, 2017: From the Homefront: https://www.surveymonkey.com/r /5RV85B5; From the Participants: https://www.surveymonkey.com/r /RPXL2Y9?sm=4uavMbpf4YADv5uu9%2bdcGA%3d%3d.

61 "It's the culture of shared stories": Monaco, interview.

61 Some 2.7 million American children: David Murphey and P. Mae Cooper, "Parents Behind Bars: What Happens to Their Children?," *ChildTrends .org*, October 2015. Cited in Nash Jenkins, "1 in 14 U.S. Children Has Had a Parent in Prison, Says New Study," *Time* magazine, October 27, 2015.

61 Nonprofits active in: These include the Women's Storybook Project of Texas (storybookproject.org), the Lutheran Social Services of Illinois (lssi.org), the Episcopal Diocese of Rochester (NY) Storybook Project (prisonministry-edr.org), and Storybook Dads UK (storybookdads.org .uk). See also Kristin Sample, "Helping Prisoners' Voices Be Heard by Their Children," *New York Times*, July 6, 2015, https://parenting.blogs .nytimes.com/2015/07/06/capturing-the-voices-of-mothers-in-prison.

61 The great-grandmother of the movement: Aunt Mary's Storybook/ Companions Journeying Together (cjtinc.org), Western Springs, IL.

61 "From the womb, the children know": Stuie Brown, phone interview by author, January 29, 2016.

63 Tempting though it may be: It is common sense that emotional bonds can be stronger or weaker; we all know this from our own lives. The fluctuating nature of attachment, and the fact that we can take steps to fortify it, inheres in such studies as Kenneth Ginsburg et al., "The Importance of Play in Promoting Healthy Child Devlopment and Maintaining Strong Parent-Child Bonds, *Pediatrics* 119, no. 1 (January 2007), http://pediatrics.aappublications.org/content/119/1/182.short; and Juulia [sic] Suvilehto et al., "Topography of Social Touching Depends on Emotional Bonds Between Humans," *Proceedings of the National Academy of Sciences of the United States of America*, November 10, 2015, https://doi .org/10.1073/pnas.1519231112.

63 "We read him board books": Danica Rommely, phone interview by author, May 25, 2016; interview by author, January 17, 2017; follow-up emails, January 19–25, 2017.

63 *Hello. . . . Meghan . . . how are you*: Gabe Rommely, Danica Rommely, and Najla, interview by author, January 17, 2017.

64 Everything had changed for the family: Colby Itkowitz, "Saying So Much Without a Sound," *Washington Post*, May 23, 2016.

64 "a drunken toddler": Itkowitz.

65 "the kindness of machines": Judith Newman, *To Siri with Love: A Mother, Her Autistic Son, and the Kindness of Machines* (New York: Harper, 2017).

Chapter 4: Turbocharging Child Development with Picture Books

68 For a brief and happy period: For a succinct explanation of this delightful phase, see "Your 5-Month-Old Baby: Learning About Object Permanence," What to Expect, updated February 27, 2015, https://www.whattoexpect .com/first-year/month-by-month/your-child-month-5.aspx.

69 Reading to children during this: Adriana Weisleder and Anne Fernald, "Talking to Children Matters: Early Language Experience Strengthens Processing and Builds Vocabulary," *Psychological Science* 24 (November 2013), https://doi.org/10.1177/0956797613488145.

69 "the nightly miracle": Shirley Jackson, *Life Among the Savages* (New York: Penguin, 1997), 129.

69 "Yet long before a baby": Pinker, *Village Effect*, 127.

71 that is the assumption: Abubakar, interview.

71 "Language comes at us": Morten Christiansen, Cornell University, Skype interview with author, April 27, 2016.

71 "mapping": Daniel Swingley, "Fast Mapping and Slow Mapping in Children's Word Learning," *Language Learning and Development* 6 (2010): 179–83, https://doi.org/10.1080/15475441.2010.484412.

71 If two adults: "A large literature indicates that talk directed to the child—rather than adult-adult or background talk—is the core data on which early language learning depends (e.g. Weisleder & Fernald, 2014)." Jessica Montag, Michael N. Jones, and Linda B. Smith, "The Words Children Hear: Picture Books and the Statistics for Language Learning," *Psychological Science* 26, no. 9 (2015), https://doi.org/10.1177/0956797615594361.

71 in a responsive way: Golinkoff et al., "(Baby) Talk to Me," 6.

71 "learn their native tongue": Golinkoff, 4.

72 the distinctive trills of its kind: Alison J. Doupe and Patricia Kuhl, "Birdsong and Human Speech: Common Themes and Mechanisms," *Annual Review of Neuroscience* 22 (1999), https://doi.org/10.1146/annurev .neuro.22.1.567. Cited in Lisa Guernsey, *Into the Minds of Babes* (New York: Basic Books, 2007), 143.

72 "If a bird doesn't hear the tutor": Quoted in Lisa Duchene, "Probing Question: How Do Songbirds Learn to Sing?," *Penn State News*, October 15, 2007, http://news.psu.edu/story/141326/2007/10/15/research/probing -question-how-do-songbirds-learn-sing.

72 "There's a lot of language learning": Christiansen, Skype interview.

72 Infants also learn from seeing: Joni N. Saby, Andrew N. Meltzoff, and Peter J. Marshall, "Infants Somatotopic Neural Responses to Seeing Human Actions: I've Got You Under My Skin," *PloS One* 8, no. 10 (2013). Also Joni N. Saby, Andrew N. Meltzoff, and Peter J. Marshall, "Neural Correlates of Being Imitated: An EEG Study in Preverbal Infants," *Social Neuroscience* 7, no. 6 (2012). Both cited in Pinker, *Village Effect*, 126–27.

73 warehousing orphans from infancy onward: For a thorough discussion of the heartbreaking plight of these children, see Charles A. Nelson, Nathan A. Fox, and Charles H. Zeanah, *Romania's Abandoned Children: Deprivation, Brain Development, and the Struggle for Recovery* (Cambridge, MA: Harvard University Press, 2014).

73 suffering from a constellation of psychological, neurological, and biological impairments: Nelson, Fox, and Zeanah, 181.

73 "When the child vocalizes": Nelson, Fox, and Zeanah, 140–41.

74 "Fun and kid-friendly iPhone applications": Cheryl Lock, "Best iPhone Apps for Babies and Toddlers," *Parents*, https://www.parents.com/fun /entertainment/gadgets/the-best-iphone-apps-for-babies-and-toddlers/.

74 "games made for your iPhone, iPad, and Android": Christen Brandt, Cheryl Lock, and Chrisanne Grise, "The Best Educational Apps for Kids," *Parents*, https://www.parents.com/fun/entertainment/gadgets/best-educational -apps-for-kids/.

74 They are no match for *us*: Steiner-Adair, *Big Disconnect*, 69, 77.

74 team at the University of Virginia: *JAMA* and Archives Journals, "Infants Do Not Appear to Learn Words from Educational DVDs." *Science Daily*, March 6, 2010, accessed April 7, 2018, www.sciencedaily.com /releases/2010/03/100301165612.htm.

74 no pedagogical value whatsoever: Judy DeLoache et al., "Do Babies Learn from Baby Media?," *Psychological Science* 21 (November 2010), https://doi .org/10.1177/0956797610384145.

75 no evidence that babies learned from the screens: Rebekah Reichert, Michael B. Robb, Jodi G. Fender, and Ellen Wartella, "Word Learning from Baby Videos," May 2010, downloaded to ArchPediatrics.com, March 28, 2011, http://cmhd.northwestern.edu/wp-content/uploads/2011/06 /Richert.Robb_.Fender.Wartella.2010.-WordLearning.pdf.

75 six to eight *fewer* new vocabulary words: University of Washington, "Baby DVDs, Videos May Hinder, Not Help, Infants' Language Development,"

Science Daily, August 8, 2007, accessed April 13, 2018, https://www.science daily.com/releases/2007/08/070808082039.htm.

75 there's not contingency": Catherine Tamis-LeMonda, interview by author, July 26, 2016.

75 technoference is a real issue for many children: McDaniel and Radesky, "Technoference."

76 For a glimpse of a child's perspective: Anna V. Sosa, "Association of the Type of Toy Used During Play with the Quantity and Quality of Parent-Infant Communication," *JAMA Pediatrics* 170 (February 2016), https://doi.org/10.1001/jamapediatrics.2015.3753.

77 These babies were indifferent to: Patricia K. Kuhl, Feng-Ming Tsao, and Huei-Mei Liu, *Proceedings of the National Academy of Sciences of the United States of America* 100 (July 22, 2003), https://doi.org/10.1073/pnas.1532872100.

77 Twelve years later: Barbara T. Conboy, Rechele Brooks, Andrew N. Meltzoff, and Patricia Kuhl, "Social Interaction in Infants' Learning of Second Language Phonetics: An Exploration of Brain-Behavior Relations," *Developmental Neuropsychology* 40 (2015), https://doi.org/10.1080/87565641.2015.1014487.

77 begin to make eye contact: Molly McElroy, "Babies' Brains Show That Social Skills Linked to Second Language Learning," *UW News*, July 27, 2015, http://www.washington.edu/news/2015/07/27/babies-brains-show-that-social-skills-linked-to-second-language-learning/.

78 Watching video excerpts: The link can be found in McElroy.

79 he had glimpsed, like Moses: Vygotsky compared himself to Moses in a private notebook shortly before he died. The quotation is cited in several places, most accessibly here: https://en.wikipedia.org/wiki/Lev_Vygotsky.

79 his writing was translated: Lev Vygotsky, *Thought and Language* (Cambridge, MA: MIT Press, 1964).

79 play is for children a crucial mechanism for self-discovery: Vygotsky, *Thought and Language*, cited in David K. Dickinson et al., "How Reading Books Fosters Language Development Around the World," *Child Development Research*, 2012, 3, https://doi.org/10.1155/2012/602807.

80 "Children benefit when": Dickinson et al., 3.

80 Fast-paced TV shows: Angeline S. Lillard and Jennifer Peterson, "The Immediate Impact of Different Types of Television on Young Children's Executive Function," *Pediatrics* 128 (May 2011), http://pediatrics.aap publications.org/content/pediatrics/early/2011/09/08/peds.2010–1919.full.pdf.

80 work done at twenty-two Head Start centers: Karen L. Bierman et al., "Promoting Academic and Social-Emotional School Readiness: The Head Start REDI Program," *Child Development* 79 (November/December 2008), https://doi.org/10.1111/j.1467–8624.2008.01227.x. The study is also helpfully summarized in "New Program Teaches Preschoolers Reading Skills, Getting Along with Others," NIH News, November 2008, https://www.nichd.nih.gov/news/releases/nov19-08-New -Program.

81 "How Reading Books Fosters": The three colleagues are Julie A. Griffith, Roberta Michnick Golinkoff, and Kathy Hirsch-Pasek.

81 "related to less aggression": Dickinson et al., "How Reading Books Fosters Language Development," 3.

81 longer periods of joint attention: Dickinson et al., 3.

81 "Consider all the ways": Dickinson et al., 6.

82 "He's just lying here and playing": Steiner-Adair, *Big Disconnect*, 70.

83 "babies are often distressed": Steiner-Adair, 71.

83 "All very successful technologies end up": Dr. Perri Klass, interview by author, October 3, 2016.

84 Theory of mind: For an accessible explanation, see Brittany N. Thompson, "Theory of Mind: Understanding Others in a Social World," *Psychology Today*, July 3, 2017, https://www.psychologytoday.com/us/blog /socioemotional-success/201707/theory-mind-understanding-others -in-social-world.

84 2015 program in the north of England: Lizzie Atkinson, "Sharing Stories, Shaping Futures: Language Development and the Shared Reading Model," *Eye* 17, no. 8 (December 2015): 41.

84 "I was reading *Solomon Crocodile*": Atkinson, 41.

86 "The wind whispers soft through the grass, hon": Adam Mansbach, *Go the F**k to Sleep*, ill. Richard Cortes (New York: Akashic, 2011), 7.

87 especially important to create a calm hiatus: Matt Wood, "Electronic Devices, Kids and Sleep: How Screen Time Keeps Them Awake," *Science Life*, University of Chicago, February 17, 2016, https://sciencelife .uchospitals.edu/2016/02/17/electronic-devices-kids-and-sleep-how -screen-time-keeps-them-awake/.

87 "Repetition and structure help children feel safe": Marie Hartwell-Walker, "The Value of a Child's Bedtime Ritual," Psychnet.com, July 17, 2016, https://psychcentral.com/lib/the-value-of-a-childs-bedtime-rou tine/.

88 "The first couple of weeks": Walter Olson and Steve Pippin, interview by author, June 13, 2016.

Chapter 5: The Rich Rewards of a Vast Vocabulary

91 a young mother named Cécile de Brunhoff made up a story: Simon Worrall, "Laurent de Brunhoff Reveals Shocking Beginning of Beloved Babar Series," *National Geographic*, December 23, 2014, https://news .nationalgeographic.com/news/2014/12/141224-babar-elephant-culture -animal-conservation-ngbooktalk/.

91 followed by six sequels by Jean de Brunhoff: The titles and dates of most of the Babar books are most easily obtained here: https://en.wikipedia.org /wiki/Babar_the_Elephant. This list omits the nine Babar books published since 2003 including the final book in the series, the surprisingly dull *Babar's Guide to Paris*. (In a December 27, 2017, review in the *Wall Street Journal*, I wrote: "For the year's most disappointing illustrations, we need look no farther than Laurent de Brunhoff's valedictory picture book, "Babar's Guide to Paris" [Abrams], a volume with illustrations so devoid of interest, so blank and lifeless, that they seem more like templates than finished drawings. Mr. de Brunhoff, who is in his 90s, has valiantly carried on the work of his father in continuing the adventures of the little elephant Babar that began in 1931, but he seems to have forgotten what used to make the books compelling. The stories were always a bit stilted, but they were saved by illustrations rich in thoughtful detail and full of dynamic modes of transport. The Babar books once teemed with cars, boats, planes, camels, elevators, elephants and hot-air balloons. At one point in this bland and inadvertently gloomy offering, Babar and Celeste dine in a featureless brasserie off empty plates and sip from unfilled glasses.")

91 a certain amount of low-level controversy: Adam Gopnik, "Freeing the Elephants: Babar Between the Exotic and the Domestic Imagination of France," in Christine Nelson, *Drawing Babar: Early Drafts and Watercolors* (New York: Morgan Library & Museum, 2008), 2–3.

94 "Words are as wild as rocky peaks": Leonard S. Marcus, *The Wand in the Word: Conversations with Writers of Fantasy* (Cambridge, MA: Candlewick Press, 2006), 74.

95 "If your attitude to language": Jon Henley, "Philip Pullman: Loosening the Chains of the Imagination," *Guardian*, August 23, 2013. Cited in Mary Roche, *Developing Children's Critical Thinking Through Picturebooks: A Guide for Primary and Early Years Students and Teachers* (London: Routledge, 2015), 56.

96 language improves a person's ability to succeed: "Reading proficiency by the third grade is the most important predictor of high school graduation and career success." Council on Early Childhood, "Literacy Promotion," 405.

96 Young children whose heads are well stocked: "Earlier age of initiation of reading aloud with a child has been shown to be associated with better preschool language skills and increased interest in reading." Council on Early Childhood, 405.

96 As neurobiologist Maryanne Wolf explains: Maryanne Wolf, *Proust and the Squid: The Story and Science of the Reading Brain* (New York: Harper Perennial, 2008), 129.

96 the Matthew effect: This term has been in broad circulation since Oakland University professor Keith Stanovich used it in his paper "Matthew Effects in Reading: Some Consequences of Individual Differences in the Acquisition of Literature," *Reading Research Quarterly*, Fall 1986, https://pdfs.semanticscholar.org/8b88/41a79b3bd90dadd5ee04df8cf7 cb63249eba.pdf. Other academic inquiries include W. B. Elley, "Vocabulary Acquisition from Listening to Stories," *Reading Research Quarterly* 24, no. 1 (2002): 174–87; and J. F. Penno et al., "Vocabulary Acquisition from Teacher Explanation and Repeated Listening to Stories: Do They Overcome the Matthew Effect?," *Journal of Educational Psychology* 94, no. 1 (2002): 23–33; both cited in Dickinson et al., "How Reading Books Fosters Language Development," 4.

96 revisit the original subject families for a 2003 report: Betty Hart and Todd R. Risley, "The Early Catastrophe: The 30 Million Word Gap by Age 3," 2003, https://www.aft.org/sites/default/files/periodicals/The EarlyCatastrophe.pdf/.

97 encouraged conversation through affirmations: Hart and Risley, *Meaningful Differences*, cited in Dickinson et al., "How Reading Books Fosters Language Development," 5.

98 researchers at Indiana University, Bloomington: Montag, Jones, and Smith, "Words Children Hear," 1489–96.

98 "shared book reading," "Unlike conversations," "a child would hear": Montag, Jones, and Smith, 1494.

99 low in socioeconomic status: Council on Early Childhood, "Literacy Promotion," 405.

99 a 2013 Stanford University study: Weisleder and Fernald, "Talking to Children Matters."

99 "When you look at the content": Tamis-LeMonda, interview.

100 "She will ask me to read": Magda Jenson, interview by author, June 2016.

102 One night the girls and I arrived at the scene: This anecdote is adapted from Meghan Cox Gurdon, "I *Love* This Story!," *NRO*, February 4, 2004, https://www.nationalreview.com/2004/02/i-love-story/.

102 Researchers at the University of Sussex: Jessica S. Horst, Kelly L. Parsons, and Natasha M. Bryan, "Get the Story Straight: Contextual Repetition

Promotes Word Learning from Storybooks," *Frontiers in Psychology* 2 (February 17, 2011), https://doi.org/10.3389/fpsyg.2011.17.

103 "Children learn vocabulary through grammar": Dickinson et al., "How Reading Books Fosters Language Development," 5.

104 "three mothers and an eggplant": P. L. Chase-Lansdale and E. Takanishi, *How Do Families Matter? Understanding How Families Strengthen Their Children's Educational Achievement* (New York: Foundation for Child Development, 2009). Cited widely, including in Dickinson et al., "How Reading Books Fosters Language Development," 5.

105 interactive reading or dialogic reading: There's a huge body of literature devoted to the power and practice of this method. To learn more, see Grover J. (Russ) Whitehurst, "Dialogic Reading: An Effective Way to Read to Preschoolers," Reading Rockets, http://www.readingrockets .org/article/dialogic-reading-effective-way-read-preschoolers; and Roche, *Developing Children's Critical Thinking*.

105 remember it far better than those who get straightforward instruction: Myae Han, Noreen Moore, Carol Vukelich, and Martha Buell, "How Play Intervention Affects the Vocabulary Learning of At-Risk Preschoolers," *American Journal of Play* 3, no. 1 (2010): 82–105.

105 "The child learns best": Roberta Michnik Golinkoff, interview by author, July 27, 2016.

106 "If you're reading with a one-year-old": Caroline Rowland, Skype interview by author, June 27, 2016.

106 The endpapers may be designed: I am grateful to Mary Roche for reminding me of this useful point in *Developing Children's Critical Thinking*, 34–39.

107 an occasional, gentle "I wonder why?": Roche, 17.

107 a skill known as auditory discrimination: Jane Fidler, interview by author, March 23, 2016.

109 even dogs have shown under MRI scanning: Karin Brulliard, "Your Dog Really Does Know What You're Saying, and a Brain Scan Shows How," *Washington Post*, August 31, 2016, https://www.washingtonpost.com /news/animalia/wp/2016/08/30/confirmed-your-dog-really-does-get-you.

110 As Jim Trelease . . . points out: "A consistent mistake made by parents and teachers is the assumption that a child's listening level is the same as his or her reading level. Until about eighth grade, that is far from true; early primary grade students listen many grades above their reading level. This means that early primary grade students are capable of hearing and understanding stories that are far more complicated than those they can read themselves." Trelease on Reading.com, http://www .trelease-on-reading.com/hey.html.

110 "the overall gist of what they are hearing or reading": E. D. Hirsch Jr., "Vocabulary Declines, with Unspeakable Results," *Wall Street Journal*, December 12, 2012.

110 "There's a hidden form of vocabulary": Doug Lemov, "Doug Lemov on Teaching," interview by Russ Roberts, *EconTalk* podcast, Library of Economics and Liberty, http://www.econtalk.org/archives/2013/12/doug_lemov_on_t.html.

112 when he's older: Dialogic reading tends naturally to fade when children are around the age of five or six and shifting, during read-aloud time, from interest chiefly in picture books to longer stories without illustration. See Dickinson, "How Reading Books Fosters Language Development," 9.

113 "'I fear yet to stir'": Bram Stoker, *Dracula* (Ware, England: Wordsworth Editions, 1993), 306.

113 "'the body of Szgany'": The term Szgany, sometimes spelled Tziganes, is no longer current. Like *gypsy*, a descriptor for the Romany people that is considered in some quarters to be retrograde and offensive.

*Chapter 6: The Power of Paying Attention—
and Flying Free*

118 "It seems that mankind is born": Sybil Marshall, *The Book of English Folktales* (New York: Overlook, 2016), 17.

118 good for the soul: James Hillman, "A Note on Story," in *Children's Literature: The Great Excluded*, vol. 3, ed. Francelia Butler and Bennett Brockman (Philadelphia: Temple University Press, 1974), https://muse.jhu.edu/article/245875. Cited in Manguel, *History of Reading*, 11.

118 "A good book is an empathy machine": Libraries Unlimited (@Libraries UnLtd), "Reading allows us to see & understand the world through the eyes of others. A good book is an empathy machine," Twitter, June 13, 2017.

118 On a recent winter day: Author visit to Park School, Baltimore, February 4, 2016.

121 Technology is training us: For a full discussion, see Adam Alter, "The Biology of Behavioral Addiction," in Alter, *Irresistible*, 68–89.

122 "You have whole generations": Simon & Schuster president and CEO Carolyn Reidy, speaking at the Frankfurt Book Fair, October 11, 2017, in remarks widely reported at the time, including by Alex Mutter, *Shelf Awareness*, October 12, 2017, http://www.shelf-awareness.com/issue.html?issue=3105#m38147.

122 tiny doses of pleasing chemicals: Alter, *Irresistible*, 68–89.

122 the attention span of the average adult: Alter, 28.

122 "the true scarce commodity": Satya Nadella, open memo to Microsoft employees, July 10, 2014. Cited widely, including by Polly Mosendz, "Microsoft's CEO Sent a 3,187 Word Memo and We Read It So You Don't Have To," *Atlantic*, https://www.theatlantic.com/technology /archive/2014/07/microsofts-ceo-sent-a-3187-word-memo-and-we-read-it -so-you-dont-have-to/374230/.

122 as Jim Trelease has pointed out: Quoted in Connie Matthiessen, "The Hidden Benefits of Reading Aloud—Even for Older Kids," GreatSchools .org, September 22, 2017, https://www.greatschools.org/gk/articles/read -aloud-to-children/.

122 Studies have uncovered a strong correlation: Megan M. McClelland et al., "Relations Between Preschool Attention Span-Persistence and Age 25 Educational Outcomes," *Early Child Research Quarterly* 28 (April 2014), https://doi.org/10.1016/j.ecresq.2012.07.008.

124 "This went on for weeks": Olson and Pippin, interview.

124 "Sometimes her voice put me to sleep": Manguel, *History of Reading*, 109–10.

124 "It can turn a child into a writer": Kate DiCamillo, "Kate DiCamillo's PSA: On the Importance of Reading Aloud to Children," YouTube, posted February 17, 2015, https://www.youtube.com/watch?v=S0c9-JM -mv00.

124 by the time kids turn five: Scholastic, *Kids & Family Reading Report*, 5th ed. (New York: Scholastic, 2015), 31–33, http://www.scholastic.com/read ingreport/Scholastic-KidsAndFamilyReadingReport-5thEdition.pdf.

125 "f u cn rd ths": For a nostalgic glimpse of the ubiquitous "School of Speed-writing" ads, see https://playingintheworldgame.com/2012/09/21/f-u-cn -rd-ths-if-you-can-read-this/.

125 "You hear it, and it's a story": Tatar, interview.

126 "Many of the references": Marcus, *Wand in the Word*, 173.

127 He was thrilled by the story: This anecdote is adapted from Gurdon, "I *Love* This Story!"

129 accused of being easy shortcuts: Remember Matthew Rubery's story about his father's friend, who apologized for claiming that he'd read a book when he had listened to it? As Rubery said, "It's viewed as cheating all the time." Rubery, interview.

131 "We were never born to read": Wolf, *Proust and the Squid*, 3.

131 Yet our brains seem not to keep close records: Rubery, interview.

131 a middle-school teacher in Wisconsin: Timothy Dolan, "The Power of Reading Aloud in Middle School Classrooms," *Education Week: Teacher*,

March 22, 2016, https://www.edweek.org/tm/articles/2016/03/22/the
-power-of-reading-aloud-in-middle.html.

132 "While I felt guilty": Michael Godsey, "The Value of Using Podcasts in Class," *Atlantic*, March 17, 2016, https://www.theatlantic.com/education/archive/2016/03/the-benefits-of-podcasts-in-class/473925/.

134 "I say, 'How did you pass high school?'": Fidler, interview.

135 "Every time my mother became pregnant": Roald Dahl, *Boy* (London: Puffin, 2010), 18–19.

136 "They gaped. They screamed": Roald Dahl, *James and the Giant Peach*, ill. Lane Smith (New York: Puffin, 1996), 42.

137 "shock effects of beauty and horror": Tatar, *Enchanted Hunters*, 29.

137 "We all like to be shocked and startled": Tatar, interview.

138 "By portraying wonderful and frightening worlds": Vigen Guroian, *Tending the Heart of Virtue: How Classic Stories Awaken a Child's Moral Imagination* (New York: Oxford University Press, 1998), 38.

138 "When we come out of a book, we're different": Jacqueline Woodson, extemporaneous remarks at Library of Congress ceremony attended by author, January 9, 2018.

138 "There is no more dynamic teacher": Zora Neale Hurston, *Dust Tracks on a Road* (New York: Harper Perennial Modern Classics, 2006), 123.

139 "Listening to Samuel Taylor Coleridge's *Kubla Khan*": Hurston, 123.

140 abolitionist and writer Fredrick Douglass: Manguel, *History of Reading*, 280–81.

140 "Slavery proved as injurious to her": Frederick Douglass, *The Narrative of the Life of Frederick Douglass*, ch. 7, http://www.pagebypagebooks.com/Frederick_Douglass/The_Narrative_of_the_Life_of_Frederick_Douglass/Chapter_VII_p1.html.

140 future missionary and preacher Thomas Johnson: Manguel, *History of Reading*, 281.

141 "Being caught reading anything": Maria Popova and Claudia Zoe Bedrick, eds., *A Velocity of Being: Letters to a Young Reader* (Brooklyn: Enchanted Lion, 2018), 58.

142 Chen was born: Chen Guangcheng, *The Barefoot Lawyer: A Blind Man's Fight for Justice and Freedom in China* (New York: Henry Holt, 2015), 15–17.

142 spent his time trapping frogs: Chen, 36, 39–42.

142 In the horror and tumult of the next decade: Stanley Karnow, *Mao and China: A Legacy of Turmoil* (New York: Penguin, 1990), 191–99.

143 "My father and I would sit under the kerosene lamp": Chen, *Barefoot Lawyer*, 43–45.

143 "The stories my father read to me": Chen, interview by author, May 17, 2017.

Chapter 7: Reading Aloud Furnishes the Mind

145 "We almost never take this out": Christine Nelson, interview by author, October 5, 2016.

145 This beautiful object: The Perrault manuscript can be seen at http://www.themorgan.org/collection/charles-perrault/manuscript.

145 possibly Perrault's son, Pierre: Morgan Library & Museum CORSAIR online collection catalog, accessed April 10, 2018, http://corsair.themorgan.org/cgi-bin/Pwebrecon.cgi?BBID=143572.

146 "the golden net-work of oral tradition": This phrase originates with the American folklorist William Wells Newall and is cited in the opening pages of Henry Louis Gates Jr. and Maria Tatar, *The Annotated African American Folktales* (New York: W. W. Norton, 2018).

147 According to Bruno Bettelheim: Bruno Bettelheim, *The Uses of Enchantment: The Meaning and Importance of Fairy Tales* (New York: Vintage Books, 2010), 168.

147 the moment of the grandmother's destruction: The tiny painting—which can be seen at http://www.themorgan.org/collection/charles-perrault/59—must, in my opinion, shows the grandmother right before the wolf eats her. After all, the grandmother is the only character in the story who is surprised in her bed. Confusingly, however, the woman in the picture wears a red cap on her head, which gives rise to the possibility that she is not in fact the grandmother but Little Red Riding Hood herself.

147 "If you want your children to be intelligent": Exuding amiable skepticism, Stephanie Winick discusses the famous quotation in "Einstein's Folklore," *Library of Congress Folklife Today*, December 18, 2013, https://blogs.loc.gov/folklife/2013/12/einsteins-folklore/.

147 "a longing for": C. S. Lewis, "On Three Ways of Writing for Children," in *Of Other Worlds* (New York: Houghton Mifflin Harcourt, 2002), 29–30.

148 "acquire a sense of horizons ": John McWhorter made this remark in the context of college students tempted to retreat into "safe spaces" rather than children who are still living at home and having stories read to them. I have borrowed it because it's elegant and it fits: a child's horizon-stretching can't start too soon! For McWhorter's full discussion, see Conor Friedersdorf, "A Columbia Professor's Critique of Campus Politics," *Atlantic*, June 30, 2017, https://www.theatlantic.com/politics/archive/2017/06/a-columbia-professors-critique-of-campus-politics/532335/.

149 "all the books on the shelves were mine": Claire Kirch, "Junot Díaz

Urges Booksellers to Walk the Talk on Diversity," *Publishers Weekly*, January 25, 2018.

150 "We all come from the past": Russell Baker, *Growing Up* (New York: Plume, 1982), 8.

150 Savage of Aveyron: Mary Losure, *Wild Boy: The Real Life of the Savage of Aveyron* (Cambridge, MA: Candlewick, 2013)

150 the terrible Tudors: Terry Deary, *Terrible Tudors and Slimy Stuarts* (New York: Scholastic, 2009).

151 "Tradition is to the community": John O'Donohue, "The Inner Landscape of Beauty," interview by Krista Tippett, *On Being*, NPR, rebroadcast August 6, 2015, https://onbeing.org/programs/john-odonohue-the-inner-landscape-of-beauty/.

151 Many of the best-known ditties: Katherine Elwes Thomas, *The Real Personages of Mother Goose* (Boston: Lothrop, Lee & Shepard, 1930).

151 embedded all over the place: Mary Roche has a good description of the phenomenon of "intertextuality," or the cross-referencing of stories and ideas, in *Developing Children's Critical Thinking*, 93–95.

152 entry point to language: This point is widely made, including in Dickinson et al., "How Reading Books Fosters Language Development," 2.

152 "Tucked inside 'Hickory, dickory'": Wolf, *Proust and the Squid*, 99.

153 "Children get a real kick": Fox, *Reading Magic*, 92–93.

153 variants of *Beauty and the Beast*: Tatar, *Beauty and the Beast*.

153 The neutrality of their characters: Bettelheim, *Uses of Enchantment*, 40.

154 "These tales are the purveyors of deep insights": Bettelheim, 26.

154 "wondrous because": Bettelheim, 19.

155 "will shine upon children with a sidewise gleam": P. L. Travers's 1943 review ran in the *New York Herald Tribune* and was featured in a 2014 exhibition of Saint-Exupéry's illustrations at the Morgan Library & Museum. It is reproduced at the end of Edward Rothstein, "70 Years on, Magic Concocted in Exile," *New York Times*, January 23, 2014, and also excerpted, with other contemporaneous reviews, in Dan Sheehan, "A Children's Fable for Adults: Saint-Exupéry's *The Little Prince*," *Literary Hub*, July 31, 2017, http://bookmarks.reviews/a-childrens-fable-for-adultsantoine-de-saint-exuperys-the-little-prince/.

156 "No one can possibly tell what tiny detail": Quoted in Tatar, *Enchanted Hunters*, 27.

156 "We think the problem is that these classics": Jack Wang, interview by author, April 25, 2016.

157 A wordless picture book gave researchers: Yana Kuchirko, Catherine S. Tamis-LeMonda, Rufan Luo, Eva Liang, "'What Happened Next?': Developmental Changes in Mothers' Questions to Children,"

Journal of Early Childhood Literacy 16 (August 17, 2015), https://doi.org /10/1177/1468798415598822.

157 "The Chinese moms tended": Tamis-LeMonda, interview.

159 A little later, in the midcentury afterglow: Leonard S. Marcus, *Golden Legacy: The Story of Golden Books* (New York: Golden Books, 2007).

160 "Making a connection to art is huge": Amy Guglielmo, interview with the author, February 16, 2016.

160 "When children have seen a painting": The Touch the Art books, originally published by Sterling, are due, as of this writing, to reappear in autumn 2019 in revamped form under the series title Peek-a-Boo Art, to be published by Orchard Books/Scholastic US.

162 "With *Rapunzel*, I was definitely showing children": Paul O. Zelinsky, interview with the author, February 10, 2016.

163 "Look at that landscape": Eik Kahng, Ellen Keiter, Katherine Roeder, David Wiesner, *David Wiesner and the Art of Wordless Storytelling* (Santa Barbara: Santa Barbara Museum of Art, distributed by Yale University Press, 2017), 17.

164 "It was a story I knew very well": Christine Rosen, interview with the author, September 12, 2016.

165 "makes your brain go tick": Jane Doonan, "Close Encounters of a Pictorial Kind," review of *Words About Pictures: The Narrative Art of Children's Books*, by Perry Nodelman, *Children's Literature Vol. 20*, 1992, accessed through Project Muse: https://muse.jhu.edu/article/246257/pdf.

165 "If we want to be able": Jonathan Cott, *There's a Mystery There: The Primal Vision of Maurice Sendak* (New York: Doubleday, 2017), 162.

166 "Interpreting pictures fully": Jane Doonan, *Looking at Pictures in Picture Books* (Stroud: Thimble Press, 1993), 8.

167 "I don't think you should do *Pride and Prejudice*": Wang, interview.

168 Accepting one of these accolades: Walter D. Edmonds, "Acceptance Paper," in *Newbery Medal Books, 1922–1955*, ed. Bertha Mahony Miller and Elinor Whitney Field (Boston: Horn Book, 1955), 223.

168 Still, we are wise not to assume: The nineteenth-century political philosopher John Stuart Mill is characteristically excellent on this point, writing in *On Liberty*, "Yet it is as evident in itself, as any amount of argument can make it, that ages are no more infallible than individuals— every age having held opinions which subsequent ages have deemed not only false but absurd; and it is as certain that many opinions, now general, will be rejected by future ages, as it is that many, once general, are rejected by the present."

168 Thomas Bowdler: Williams, *Social Life of Books*, 178–80.

169 small Chicago publisher: NewSouth Books, which published the bowd-

lerized editions of Twain, defended its editorial choices shortly before
the books came out in early 2011: "We saw the value in an edition that
would help the works find new readers. If the publication sparks good
debate about how language impacts learning or about the nature of censorship or the way in which racial slurs exercise their baneful influence,
then our mission in publishing this new edition of Twain's works will be
more emphatically fulfilled."

169 "Twain used the 'n-word'": Amy Guth, "Epithets Edited out of 'Tom
Sawyer,' 'Finn,'" updated by Courtney Crowder, *Chicago Tribune*, January 5, 2011.

169 "How to acknowledge an author's darker side": Miller, *Magician's
Book*, 171.

170 a notorious scene: Laura Ingalls Wilder, *Little Town on the Prairie*, illustrator Garth Williams (New York: Harper Trophy, 1994), 257–59.

171 in an episode: *Bewitched*: season 2, episode 26, "Baby's First Paragraph,"
March 1966, https://www.imdb.com/title/tt0523058/.

172 "Great art has often": Camille Paglia, "Camille Paglia on Movies, #MeToo
and Modern Sexuality: 'Endless, Bitter Rancor Lies Ahead,'" *Hollywood
Reporter*, February 27, 2018, https://www.hollywoodreporter.com/news
/camille-paglia-movies-metoo-modern-sexuality-endless-bitter-rancor
-lies-1088450.

173 "History, after all, is people": Elizabeth Janet Gray, "Acceptance Paper,"
in Miller and Field, *Newbery Medal Books*, 240.

173 "gives us a profound sense": Gray, 239.

173 Louise Erdrich's *The Birchbark House*: I am grateful to Bruce Handy for
reminding me of this wonderful book, the first in a series. Louise Erdrich, *The Birchbark House* (Los Angeles: Disney/Hyperion, 2002).

174 "If all mankind minus one": John Stuart Mill, *On Liberty* (Indianapolis:
Bobbs-Merrill, 1982), 21.

174 One of the best models: Virgil M. Hillyer, *A Child's History of the World*
(Hunt Valley, MD: Calvert Education Services, 1997), 163–65.

Chapter 8: From the Nursery to the Nursing Home: Why Reading Aloud Never Gets Old

178 "It is really hard to sit": Linda Khan, conversation with the author, December 2017.

178 "When I was a kid [my mother] would read to me": Michelle Homer,
Deborah Duncan, "Emotional Neil Bush on his mother's life and legacy,"
KHOU, April 16, 2018, https://www.khou.com/article/news/community
/emotional-neil-bush-on-his-mothers-life-and-legacy/285-540211907.

179 a charming account in the *New Yorker*: Niccolo Tucci, "The Great For-
eigner," *New Yorker*, November 22, 1947, https://www.newyorker.com
/magazine/1947/11/22/the-great-foreigner.

179 "I believe that one of the strongest motives": Albert Einstein, "Principles of Research," a celebrated toast that Einstein delivered in honor of
Max Planck's 60th birthday celebration, http://www.neurohackers.com
/index.php/fr/menu-top-neurotheque/68-cat-nh-spirituality/99-principles
-of-research-by-albert-einstein.

181 an unpleasant surprise: Lauri Hornik, conversation with author, January
9, 2018; follow-up phone interview, January 26, 2018.

182 "Literature must be taken and broken to bits": Vladimir Nabokov,
"Nabokov on Dostoyevsky," *New York Times Magazine*, 1981, https://www
.nytimes.com/1981/08/23/magazine/nabokov-on-dostoyevsky.html.

182 One June afternoon: The reading group was held at the Michael Sobell
Jewish Community Center, Golders Green Road, London.

187 "rich, varied, non-prescriptive diet of serious literature": Josie Billington et al., "An Investigation into the Therapeutic Benefits of Reading in
Relation to Depression and Well-Being," University of Liverpool, 2010,
https://www.thepilgrimtrust.org.uk/wp-content/uploads/MERSEY
BEAT-Executive-Summary.pdf.

187 feelings of relaxation: In an interesting unrelated study, below, researchers teased apart the social-emotional effects of fiction and nonfiction,
finding that "exposure to fiction was positively correlated with social
support. Exposure to nonfiction, in contrast, was associated with loneliness, and negatively related to social support." Raymond Mar, Keith
Oatley, and Jordan B. Peterson, "Exploring the Link Between Reading
Fiction and Empathy: Ruling Out Individual Differences and Exam-
ining Outcomes," *Communications* 34, no. 4 (2009): 407–28, doi: 10.1515
/COMM.2009.025.

187 "This effect is likely because books engage": Avni Bavishi, Martin D.
Slade, and Becca Levy, "A Chapter a Day: Association of Book Reading
with Longevity," *Social Science & Medicine* 164 (September 2016): 44–48.

187 A 2017 paper from clinicians: Josie Billington et al., "A Literature-Based
Intervention for Older People Living with Dementia," evaluation report, Centre for Research into Reading, Literature and Society, University of Liverpool, August 15, 2017, https://issuu.com/emmawalsh89/docs
/a-literature-based-intervention-for.

188 "Experience and use of language do matter": Christiansen, Skype conversation.

188 sharpen cognitive skills: Rui Nouchi et al., "Reading Aloud and Solving Simple Arithmetic Calculation Intervention (Learning Therapy)

Improves Inhibition, Verbal Episodic Memory, Focus Attention and Processing Speed in Healthy Elderly People: Evidence from a Randomized Controlled Trial," *Frontiers in Human Neuroscience* 10 (May 17, 2016), doi:10.3389/fnhum.2016.00217.

188 "It is almost as if literature 'raises the bar'": "Connect Realise Change: The Reader," https://vdocuments.site/connect-realise-change-high -quality.html.

188 loneliness as a cultural phenomenon: Vivek Murthy, "Work and the Loneliness Epidemic," *Harvard Business Review*, September 2017.

189 Lonely hearts face double the mortality risk: European Society of Cardiology Press Office, "Loneliness is Bad for the Heart," June 9, 2018, https://www.escardio.org/The-ESC/Press-Office/Press-releases/loneli ness-is-bad-for-the-heart?hit=wireek.

189 Dogs do, too: Andy Newman, "How to Heal a Traumatized Dog: Read It a Story," *New York Times*, June 9, 2016, http://www.nytimes.com /2016/06/12/nyregion/how-to-heal-a-traumatized-dog-read-it-a-story .html.

189 "Ten or fifteen years ago": Victoria Wells, ASPCA, interview by author, October 3, 2016.

191 "You read something which you thought": James Baldwin, *James Baldwin: The Last Interview and Other Conversations* (Hoboken: Melville House, 2014).

191 Antyllus thought it was: Manguel, *History of Reading*, 60–61.

191 He had the fanciful idea: Aline Rousselle, *Porneia: On Desire and the Body in Antiquity* (Eugene, OR: Wipf and Stock, 2013), 11.

Chapter 9: There Is No Present Like the Time

195 a middle-class family, the Rashids: Julie and Alex Rashid and their children, interviews by author, May 29 and September 11, 2016.

198 A 2017 study, "Learning on Hold": Jessa Reed et al., "Learning on Hold: Cellphones Sidetrack Parent-Child Interactions," *Developmental Psychology* 53 (August 2017), https://www.ncbi.nlm.nih.gov/pubmed/28650177.

199 "Even with technology in our lives": Klass, interview.

199 we don't know exactly what influence: Abubakar, interview.

200 "There are few things that feel to a person": LeVar Burton, "Levar Burton Reads to You," interview with Lauren Ober, *The Big Listen*, NPR, November 23, 2017, https://biglisten.org/shows/2017-11-23/levar-burton -reads.

201 In a nanosecond: True story! This took place late in the afternoon of March 30, 2016.

202 "This memory made my family realize": Annie Holmquist, email exchange with author, August 2, 2017.

203 "I think it is *great*": Carolyn Siciliano, interview by author, June 5, 2016.

204 a subtle technique: Alter, *Irresistible*, 272–73.

206 As a fellow enthusiast said: It was Walter Olson.

206 Researchers at NYU have observed: Tamis-LeMonda, interview.

209 developmental psychologists who argue: Pinker, *Village Effect*, 181.

210 "I ask myself, why did I stop": Amelia DePaolo, interview by author, July 20, 2017.

210 "I said to her, 'I was thinking'": Lauri Hornik, interview by author, January 26, 2018.

210 a quaint volume: Walter Taylor Field, *Fingerposts to Children's Reading* (Chicago: A. C. McClurg, 1907), cited in Annie Holmquist, *Intellectual Takeout*, November 9, 2015, http://www.intellectualtakeout.org/blog/5 -tips-1907-turning-your-child-reader.

211 "a special time with their parents": Of children ages six to seventeen, 78 percent chose "It is/was a special time with my parent" as their top reason for why they enjoyed being read to. Scholastic, *Kids & Family Reading Report*, 5th ed., 35.

211 "need to feel competent at what they do": Sebastian Junger, *Tribe: On Homecoming and Belonging* (New York: Twelve, 2016), 22.

Afterword

217 "It is only with the heart": Antoine de Saint-Exupéry, *The Little Prince* (London: Egmont, 2012), 68.

217 "astonished by a sudden understanding": Saint-Exupéry, 73–74.

218 When the writer and illustrator Anna Dewdney: Shannon Maughn, "Obituary: Anna Dewdney," *Publishers Weekly*, September 6, 2016, http:// publishersweekly.tumblr.com/post/150056066571/childrens-author -illustrator-and-educator-anna.

218 "When we read with a child": Anna Dewdney, "How Books Can Teach Your Child to Care," *Wall Street Journal*, August 7, 2013, https://blogs .wsj.com/speakeasy/2013/08/07/why-reading-to-children-is-crucial-not -just-for-literacy/.

READ-ALOUD BOOKS MENTIONED
IN *THE ENCHANTED HOUR*

—————

All highly recommended!

20,000 Leagues Under the Sea, by Jules Verne

Abel's Island, by William Steig

The Absolutely True Diary of a Part-Time Indian, by Sherman Alexie

Adam of the Road, by Elizabeth Gray, illustrated by Robert Lawson

The Adventures of Huckleberry Finn, by Mark Twain

Alice's Adventures in Wonderland, by Lewis Carroll, illustrated by John Tenniel

The Amazing Bone, by William Steig

Andrew's Loose Tooth, by Robert Munsch

Around the World with Ant and Bee, by Angela Banner

Art & Max, by David Wiesner

Art Up Close, by Claire d'Harcourt

The BabyLit Series, by Jennifer Adams, illustrated by Alison Oliver

A Baby Sister for Frances, by Russell Hoban, illustrated by Lillian Hoban

The Bear Ate Your Sandwich, by Julia Sarcone-Roach

Beowulf, retold by Michael Morpurgo, illustrated by Michael Foreman

The BFG, by Roald Dahl, illustrated by Quentin Blake

The Big Honey Hunt, by Stan and Jan Berenstain

The Birchbark House, by Louise Erdrich

Boy, by Roald Dahl, illustrated by Quentin Blake

Brush of the Gods, by Lenore Look, illustrated by Meilo So

The Bunny Book, by Patsy Scarry, illustrated by Richard Scarry

A Butterfly Is Patient, by Dianna Hutts Aston, illustrated by Sylvia Long

Charlie and the Chocolate Factory, by Roald Dahl, illustrated by Quentin Blake

Charlotte's Web, by E. B. White, illustrated by Garth Williams

A Child's Book of Art: Great Pictures, First Words, by Lucy Micklethwait

A Child's History of the World, by Virgil Hillyer, illustrated by Carle Michel Boog and M. S. Wright

The Chronicles of Narnia, by C. S. Lewis

The Cozy Classics Series, by Jack Wang and Holman Wang
A Cricket in Times Square, by George Selden
Curious George, by H. A. Rey and Margret Rey
D'Aulaires' Book of Greek Myths, by Ingri d'Aulaire and Edgar Parin d'Aulaire
The Day the Crayons Quit, by Drew Daywalt, illustrated by Oliver Jeffers
Dear Zoo, by Rod Campbell
Demolition, by Sally Sutton, illustrated by Brian Lovelock
Dominic, by William Steig
The Emperor's New Clothes, by Hans Christian Andersen, illustrated by
    Virginia Lee Burton
Fahrenheit 451, by Ray Bradbury
Farmer Boy, by Laura Ingalls Wilder, illustrated by Garth Williams
The Fault in Our Stars, by John Green
The Fellowship of the Ring, by J. R. R. Tolkien
Finding Winnie, by Lindsay Mattick, illustrated by Sophie Blackall
The Fire Station, by Robert Munsch
Flotsam, by David Wiesner
Fortune, by Diane Stanley
Frog, Where Are You?, by Mercer Mayer
The Giver, by Lois Lowry
Go, Dog. Go!, by P. D. Eastman
Good Night, Gorilla, by Peggy Rathmann
Goodnight iPad, by "Ann Droyd," a.k.a. David Milgrim
Goodnight Moon, by Margaret Wise Brown, illustrated by Clement Hurd
Green Eggs and Ham, by Dr. Seuss
Grimm's Fairy Tales, by Jacob and Wilhelm Grimm
The Gruffalo, by Julia Donaldson, illustrated by Axel Scheffler
Guess How Much I Love You, by Sam McBratney, illustrated by Anita Jeram
The Happy Lion, by Louise Fatio, illustrated by Roger Duvoisin
Harry Potter and the Sorcerer's Stone, by J. K. Rowling, illustrated by Mary
    GrandPré
His Royal Highness, King Baby, by Sally Lloyd-Jones, illustrated by David Roberts
The Hobbit, by J. R. R. Tolkien
Holes, by Louis Sachar
The House of Sixty Fathers, by Meindert DeJong, illustrated by Maurice Sendak
The Hunger Games, by Suzanne Collins
I Love You to the Moon and Back, by Amelia Hepworth, illustrated by Tim Warnes
The Iliad, retold by Gillian Cross, illustrated by Neil Packer
The Island of the Blue Dolphins, by Scott O'Dell
Jacob Two-Two Meets the Hooded Fang, by Mordecai Richler, illustrated by Dušan
    Petričić

Pippi Goes on Board, by Astrid Lindgren, illustrated by Florence Lamborn
Rapunzel, by Paul O. Zelinsky
Redwall, by Brian Jacques
Rip Van Winkle, by Washington Irving
Rumpelstiltskin, by Paul O. Zelinsky
Sam and Dave Dig a Hole, by Mac Barnett, illustrated by Jon Klassen
The Sand Castle Contest, by Robert Munsch
The Secret Garden, by Frances Hodgson Burnett, illustrated by Tasha Tudor
The Seven Wise Princesses, by Wafa' Tarnowska, illustrated by Nilesh Mistry
The Shadow, by Donna Diamond
Sleepy Solar System, by John Hutton, illustrated by Doug Cenko
The Snowy Day, by Ezra Jack Keats
Solomon Crocodile, by Catherine Rayner
The Story About Ping, by Marjorie Flack, illustrated by Kurt Wiese
The Story of Babar, by Jean de Brunhoff
The Story of Ferdinand, by Munro Leaf, illustrated by Robert Lawson
Struwwelpeter, by Heinrich Hoffman
Stuart Little, by E. B. White, illustrated by Garth Williams
The Swiss Family Robinson, by Johann Wyss
Sylvester and the Magic Pebble, by William Steig
The Tale of Mrs. Tiggy-Winkle, by Beatrix Potter
The Tales of Mother Goose, by Charles Perrault, illustrated by Gustave Doré
Tawny Scrawny Lion, by Katherine Jackson, illustrated by Gustaf Tenggren
This Is Not My Hat, by Jon Klassen
Through the Looking Glass, "Jabberwocky," by Lewis Carroll, illustrated by Sir
    John Tenniel
The Tinderbox, retold by Stephen Mitchell, illustrated by Bagram Ibatoulline
The Touch the Art Series and *Brush Mona Lisa's Hair*, by Amy Guglielmo and
    Julie Appel
The Travels of Babar, by Jean de Brunhoff
Treasure Island, by Robert Louis Stevenson
Turtles All the Way Down, by John Green
The Underneath, by Kathi Appelt, illustrated by David Small
Utterly Lovely One, by Mary Murphy
The Wind in the Willows, by Kenneth Grahame, illustrated by Arthur Rackham
Winnie-the-Pooh, by A. A. Milne, illustrated by Ernest H. Shepard
The Wolves of Willoughby Chase, by Joan Aiken, illustrated by Pat Marriott
The Wonderful World of Oz, by L. Frank Baum, illustrated by W. W. Denslow
A Wrinkle in Time, by Madeleine L'Engle
Young Titan: The Making of Winston Churchill, by Michael Shelden

MORE SUGGESTED STORIES
FOR READING ALOUD

•——•

Adventures in the Wide World

Black Ships Before Troy, by Rosemary Sutcliff, illustrated by Alan Lee
Clever Ali, by Nancy Farmer, illustrated by Gail de Marcken
Cloud Tea Monkeys, by Mal Peet and Elspeth Graham, illustrated by Juan
    Wijngaard
How I Learned Geography, by Uri Shulevitz
Night Sky Dragons, by Mal Peet and Elspeth Graham, illustrated by Patrick
    Benson
Pippi Longstocking, by Astrid Lindgren
Rosie's Magic Horse, by Russell Hoban, illustrated by Quentin Blake

Artful Books

The Art Book for Children, volumes 1 & 2, by the editors of Phaidon Press
Babar's Museum of Art, by Laurent de Brunhoff
Blue Rider, by Geraldo Valério
The First Drawing, by Mordicai Gerstein
Imagine!, by Raúl Colón
Lives of the Great Artists, by Charlie Ayers
Old Masters Rock, by Maria-Christina Sayn-Wittgenstein Nottebohm

Bedtime

And If the Moon Could Talk, by Kate Banks, illustrated by Georg Hallensleben
The Bear in the Book, by Kate Banks, illustrated by Georg Hallensleben
The Big Red Barn, by Margaret Wise Brown, illustrated by Felicia Bond
The Prince Won't Go to Bed!, by Dayle Ann Dodds, illustrated by Krysten
    Brooker

Power Down, Little Robot, by Anna Staniszewski, illustrated by Tim Zeltner
What Can You Do with a Shoe?, by Beatrice de Regniers, illustrated by Maurice
    Sendak
Where's My Teddy?, by Jez Alborough

Being Kind

Gorilla! Gorilla!, by Jeanne Willis, illustrated by Tony Ross
The Hundred Dresses, by Eleanor Estes, illustrated by Louis Slobodkin
I Walk with Vanessa, by Kerascoët
Last Stop on Market Street, by Matt de la Peña, illustrated by Christian Robinson
Stella's Starliner, by Rosemary Wells
The Wild Girl, by Chris Wormell
Wolf in the Snow, by Matthew Cordell

Being Small

Du Iz Tak?, by Carson Ellis
Emily's Balloon, by Komako Sakai
Edward's First Night Away, by Rosemary Wells
King Jack and the Dragon, by Peter Bently, illustrated by Helen Oxenbury
Little Wolf's First Howling, by Laura McGee Kvasnowsky and Kate Harvey
    McGee
The Tilly and Friends books, by Polly Dunbar
Waiting, by Kevin Henkes

Cars and Trucks and Things That Go Vroom

20 Big Trucks in the Middle of the Street, by Mark Lee, illustrated by Kurt Cyrus
Demolition, by Sally Sutton, illustrated by Brian Lovelock
Fire Truck, by Peter Sís
Here Comes the Train, by Charlotte Voake
Machines Go to Work, by William Low
Machines Go to Work in the City, by William Low
The Caboose Who Got Loose, by Bill Peet

Concept Books, illustrated by Counting, Colors, Opposites

Alphablock, by Christopher Franceschelli, illustrated by Peskimo
Before After, by Anne-Margot Ramstein & Mathias Arégui
Cockatoos, by Quentin Blake

Ducks Away!, by Mem Fox, illustrated by Judy Horacek
Llamaphones, by Janik Coat
My Pictures After the Storm, by Éric Veillé
Opposites, by Xavier Deneux

Embracing Eccentricity

Cecil the Pet Glacier, by Matthea Harvey, illustrated by Giselle Potter
Henry Huggins, by Beverly Cleary, illustrated by Louis Darling
How Tom Beat Captain Najork and His Hired Sportsmen, by Russell Hoban,
    illustrated by Quentin Blake
Imogene's Antlers, by David Small
Marshall Armstrong Is New to Our School, by David Mackintosh
Martha Speaks, by Susan Meddaugh
Traction Man Meets Turbo Dog, by Mini Grey

Fables & Just Desserts

The Chinese Emperor's New Clothes, by Ying Chang Compestine, illustrated by
    David Roberts
The Goat-Faced Girl, by Leah Marinsky Sharpe, illustrated by Jane Marinsky
Hubert's Hair-Raising Adventure, by Bill Peet
Just So Stories, by Rudyard Kipling
Lousy Rotten Stinkin' Grapes, by Margie Palatini, illustrated by Barry Moser
The Storyteller, by Evan Turk
The Wolf, the Duck, and the Mouse, by Mac Barnett, illustrated by Jon Klassen

Fairy and Folk Tales, Old and New

The Bearskinner, by Laura Amy Schlitz, illustrated by Max Grafe
Brave Red, Smart Frog, by Emily Jenkins, illustrated by Rohan Daniel Eason
The Girl with a Brave Heart, by Rita Jahanforuz, illustrated by Vali Mintzi
Iron Hans, by Stephen Mitchell, illustrated by Matt Tavares
The Jungle Grapevine, by Alex Beard
Robin Hood, by David Calcutt, illustrated by Grahame Baker-Smith
The White Elephant, by Sid Fleishman, illustrated by Robert McGuire

Family Life

Blackout, by John Rocco
Building Our House, by Jonathan Bean

Leave Me Alone!, by Vera Brosgol
The Money We'll Save, by Brock Cole
Rotten Ralph, by Jack Gantos, illustrated by Nicole Rubel
Toys Go Out and *Toys Come Home* and *Toy Dance Party*, by Emily Jenkins, illustrated by Paul O. Zelinsky
Yard Sale, by Eve Bunting, illustrated by Lauren Castillo

Finding One's Place

The Obvious Elephant, by Bruce Robinson, illustrated by Sophie Windham
Wee Gillis, by Munro Leaf, illustrated by Robert Lawson
Leon the Chameleon, by Mélanie Watt
Noodle, by Munro Leaf, illustrated by Ludwig Bemelmans
Make Way for Ducklings, by Robert McCloskey
Piper, by Emma Chichester Clark
Library Lion, by Michelle Knudsen, illustrated by Kevin Hawkes

Friendship and Love

The Chirri & Chirra books, by Kaya Doi, translated from the Japanese by Yuki Kaneko
The Happy Lion Roars, by Louise Fatio, illustrated by Roger Duvoisin
Martin Pebble, by Jean-Jacques Sempé
Paul Meets Bernadette, by Rosy Lamb
The Reluctant Dragon, by Kenneth Grahame, illustrated by Ernest H. Shepard
The Runaway Bunny, by Margaret Wise Brown, illustrated by Clement Hurd
The Song of Delphine, by Kenneth Kraegel

Heavy Treasures

The American Story, by Jennifer Armstrong, illustrated by Roger Roth
The Annotated African American Folktales, by Maria Tatar with Henry Louis Gates Jr.
The Brambly Hedge Treasury, by Jill Barklem
Complete Adventures of Curious George, by H. A. Rey and Margret Rey
Maps, by Aleksandra Mizielinska and Daniel Mizielinski
Norse Myths: Tales of Odin, Thor and Loki, by Kevin Crossley-Holland, illustrated by Jeffrey Alan Love
The World of Robert McCloskey, by Robert McCloskey

Ingenuity and Enterprise

The Arabian Nights, by Wafa' Tarnowska, illustrated by Carole Henaff

Anatole and *Anatole and the Cat*, by Eve Titus, illustrated by Paul Galdone

The Donut Chef, by Bob Staake

The Duchess Bakes a Cake, by Virginia Kahl

The Dunderheads, by Paul Fleischman, illustrated by David Roberts

Good Enough to Eat, by Brock Cole

Stone Soup, by Marcia Brown

Lengthier Stories for Young Listeners

Ash Road, by Ivan Southall

A Drowned Maiden's Hair, by Laura Amy Schlitz

The Graveyard Book, by Neal Gaiman, illustrated by Dave McKean

Half Magic and *Magic by the Lake*, by Edward Eager, illustrated by N. M. Bodecker

The Inquisitor's Tale, by Adam Gitwitz, illustrated by Hatem Aly

The Miraculous Journey of Edward Tulane, by Kate DiCamillo, illustrated by Bagram Ibatoulline

A Tale Dark and Grimm, by Adam Gidwitz, illustrated by Dan Santat

Poetry and Rhymes

17 Kings and 42 Elephants, by Margaret Mahy, illustrated by Patricia MacCarthy

Beastly Verse, by JooHee Yoon

A Child's Garden of Verses, by Robert Louis Stevenson, illustrated by Tasha Tudor

Life Doesn't Frighten Me, by Maya Angelou, illustrated by paintings by Jean-Michel Basquiat

Orange Pear Apple Bear, by Emily Gravett

They All Saw a Cat, by Brendan Wenzel

What the Ladybug Heard, by Julia Donaldson, illustrated by Lydia Monks

Seek-and-Find

Fish on a Walk, by Eva Muggenthaler

In the Town All Year 'Round, by Rotraut Susanne Berner

Lost and Found: Adèle and Simon in China, by Barbara McClintock

One Is Not a Pair, by Britta Tekkentrup

Undercover, by Bastien Contraire

Where's Walrus?, by Stephen Savage

Where's Warhol?, by Catherine Ingram, illustrated by Andrew Rae

Stories from the Real World

Dadblamed Union Army Cow, by Susan Fletcher, illustrated by Kimberly Bulcken Root

Josephine: The Dazzling Life of Josephine Baker, by Patricia Hruby Powell, illustrated by Christian Robinson

Shackleton's Journey and *The Wolves of Currumpaw*, by William Grill

Sky Boys: How They Built the Empire State Building, by Deborah Hopkinson, illustrated by James E. Ransome

Strong Man: The Story of Charles Atlas, by Meghan McCarthy

Tiny Creatures: The World of Microbes, by Nicola Davies, illustrated by Emily Sutton

Wise Guy: The Life and Philosophy of Socrates, by M. D. Usher, illustrated by William Bramhall

Strong Feelings

Brave Martha, by Margot Apple

The Funeral, by Matt James

Gorilla, by Anthony Browne

Grumpy Bird, by Jeremy Tankard

Jabari Jumps, by Gaia Cornwall

Maybe a Bear Ate It!, by Robie H. Harris, illustrated by Michael Emberley

The Terrible Plop, by Ursula Dubosarsky, illustrated by Andrew Joyner

Witty Wordplay

The Alphabet Thief, by Bill Richardson, illustrated by Roxana Bikadoroff

Bashful Bob and Doleful Dorinda, by Margaret Atwood, illustrated by Dušan Petričić

Betty's Burgled Bakery, by Travis Nichols

Mirror Mirror, by Marilyn Singer, illustrated by Jośee Masse

Mom and Dad Are Palindromes, by Mark Shulman, illustrated by Adam McCauley

Stegothesaurus, by Bridget Heos, illustrated by T. L. McBeth

Take Away the A and *Where's the Baboon?*, by Michaël Escoffier, illustrated by Kris di Giacomo

Wordless Picture Books

The Chicken Thief and *Fox and Hen Together*, by Béatrice Rodriguez
The Farmer and the Clown, by Marla Frazee
The Hero of Little Street, by Gregory Rogers
Ice, by Arthur Geisert
Journey and *Quest* and *Return*, by Aaron Becker
The Red Book and *Red Again*, by Barbara Lehman
Wave, by Suzy Lee

FOR OLDER LISTENERS

Classic Stories

"The Body Snatcher," by Robert Louis Stevenson
"Green Sealing Wax," by Colette
"The Ice Maiden," by Hans Christian Andersen
"The Legend of Sleepy Hollow," by Washington Irving
"The Lumber Room," by Saki
"The Necklace," by Guy de Maupassant
A Study in Scarlet, by Arthur Conan Doyle
"The Tell-Tale Heart," by Edgar Allan Poe
"To Build a Fire," by Jack London

Classic Novels and Novellas

Animal Farm, by George Orwell
A Christmas Carol, by Charles Dickens
The Code of the Woosters, by P. G. Wodehouse
The Death of Ivan Ilych, by Leo Tolstoy
Dr. Jekyll and Mr. Hyde, by Robert Louis Stevenson
Emma, by Jane Austen
The Great Gatsby, by F. Scott Fitzgerald
The Haunting of Hill House, by Shirley Jackson
True Grit, by Charles Portis

INDEX

ABOUT THE AUTHOR

MEGHAN COX GURDON is an essayist, book critic, and former foreign correspondent who has been the *Wall Street Journal*'s children's book reviewer since 2005. Her work has appeared widely, in publications such as the *Washington Examiner*, the *Daily Telegraph*, the *Christian Science Monitor*, the *Washington Post*, the *San Francisco Chronicle*, and *National Review*. A graduate of Bowdoin College, she lives in Bethesda, Maryland, with her husband, Hugo Gurdon, and their five children.